Encyclopaedia of Mathematical Sciences
Volume 134

Invariant Theory and Algebraic Transformation Groups V

Lex E. Renner

Linear Algebraic Monoids

Springer

Author

Lex E. Renner
Department of Mathematics
University of Western Ontario
London, Ontario
Canada N6A 5B7
e-mail: lex@uwo.ca

Mathematics Subject Classification (2000):
06A07, 14L30, 14L40, 14R20, 16W30, 20G05, 20G40, 20M17, 20M20

ISSN 0938-0396
ISBN 3-540-24241-4 Springer Berlin Heidelberg New York

Springer is a part of Springer Science+Business Media GmbH
springeronline.com

Printed in The Netherlands

Typesetting: by the author using a Springer LaTeX macro package
Production: LE-TeX Jelonek, Schmidt & Vöckler GbR, Leipzig
Cover Design: E. Kirchner, Heidelberg, Germany
Printed on acid-free paper 46/3142 YL 5 4 3 2 1 0

To my parents, Roy and Jo.

Preface

The object of this monograph is to document what is most interesting about linear monoids. We show how these results fit together into a coherent blend of semigroup theory, groups with BN-pair, representation theory, convex geometry and algebraic group theory. The intended reader is one who is familiar with some of these topics, and is willing to learn about the others.

The intention of the author is to convince the reader that reductive monoids are among the darlings of algebra. We do this by systematically assembling many of the major known results with many proofs, examples and explanations. To further entice the reader, we have included many exercises.

The theory of linear algebraic monoids is quite recent, originating around 1980. Both Mohan Putcha and the author began the systematic study independently. But this development would not have been possible without the pioneering work of Chevalley, Borel and Tits on algebraic groups. Also, there is the related, but more general theory of spherical embeddings, developed largely by Brion, Luna and Vust. These theories were developed somewhat independently, but it is always a good idea to interpret monoid results in the combinatorial apparatus of spherical embeddings.

Each chapter of this monograph is focussed on one or more of the major themes of the subject. These are: classification, orbits, geometry, representations, universal constructions and combinatorics. There is an inherent diversity and richness in the subject that usually rewards a stalwart investigation.

I would like to acknowledge some of those whose efforts or participation have made this monograph possible. The late Roy R. Douglas, my Ph. D. supervisor, whose boundless, open-minded enthusiasm got me started on the study of algebraic monoids. Mohan S. Putcha, for often taking the next step when I was stuck. My former students Wenxue Huang, Zhuo Li and Zhenheng Li, for suggesting improvements and helping me not to forget how mathematical ideas move from one generation to the next. Lou Solomon, for finding the fundamental links with combinatorics and Hecke-Iwahori algebra. Karl Hofmann and Èrnest Vinberg, for giving me the opportunity to assess and present my ideas in the broader context of Positivity in Lie Theory. Vladimir Popov, who invited me into this exciting EMS project with Springer-Verlag.

London, Canada,
July, 2004

Lex E. Renner

Contents

1 Introduction

The theory of linear algebraic monoids has been developed significantly only over the last twenty-five years, due largely to the efforts of Putcha and the author. It culminates a natural blend of algebraic groups, torus embeddings and semigroups. Unfortunately, this work had not been made as accessible as it might have been. Many of the fundamental developments were obtained after Putcha published his basic monograph "Linear algebraic monoids" in 1988. Solomon's 1995 survey "An introduction to reductive monoids" provides an engaging introduction to the theory of reductive monoids for a reader with an interest in algebra and combinatorics, but without requiring a lot of background from semigroups and algebraic group theory.

The purpose of this monograph is to update the literature with a detailed survey of the latest developments, along with many proofs, examples and explanations. At the same time, we hope to make the discussion reasonably self contained, even though the prerequisites are quite high. Our hope is that we can make this subject, and its methods, more accessible to a larger audience.

The systematic development of the theory of algebraic monoids began around 1978. Both Putcha and the author independently saw the potential in these monoids for a rich, and highly structured blend of group theory, combinatorics and torus embeddings. Putcha began his investigation around 1978 by experimenting with some of the main ideas of semigroup theory: Green's relations, regularity, semilattices, and so on. His efforts yielded a lot of technically useful information. In particular, he established control of the idempotent set of an irreducible monoid.

About the same time I started writing my Ph.D. thesis armed with some encouraging success in applications to rational homotopy theory. These applications inspired the hope that reductive monoids could be properly understood as a geometric blend of the Zariski closure of a maximal, split torus, and the unit group.

The first major result came around 1982. Reductive monoids are regular. At this point we knew for certain that we were onto something special. Any regular monoid is inevitably (somehow) determined by its unit group and its idempotent set. A major developmental theme from this point on was the rôle of the idempotent set of any irreducible monoid. From there, Putcha found his *cross section lattice* Λ, the most useful way to control the $G \times G$-orbits of M. Armed with Λ, and Grosshans' *codimension 2 condition*, it was then possible for me to develop the classification theory of reductive monoids in a way that allowed a description of the set of morphisms from any reductive normal monoid. About that time I started my investigation of the analogue of the Bruhat decomposition for reductive monoids.

Around 1986, Putcha observed that each reductive monoid M has a *type map*, $\lambda : \Lambda \to 2^S$. This is truly the monoid analogue of the Dynkin diagram: it determines M up to a kind of central extension, it determines Nambooripad's biordered set of idempotents, and it determines the set of $B \times B$-orbits of M. This is just what Putcha needed to develop his abstract theory of *Monoids of Lie type*, the monoid analogue of the theory of groups with BN pair. But it is no soft excercise in generalization theory (see Chapter 10). In any case, the type map is the exact, minimal, discrete entity that can be used to determine the salient structure of a Monoid of Lie type.

In a joint effort, around 1988, we determined explicitly a large class of type maps. These are the type maps of $\mathfrak{J}$-irreducible monoids. A reductive monoid M is *$\mathfrak{J}$-irreducible* if it has exactly one, non zero, minimal $G \times G$-orbit. This leads to some speculation about what is possible in general. On the one hand, it is impossible to list all type maps but, on the other hand, there are still some interesting questions here. We have recently determined the type maps of reductive monoids with exactly two minimal, non zero $G \times G$-orbits.

In another joint effort, arround 1990, we investigated the irreducible, modular representations of a finite monoid of Lie type. By combining the results of semigroup representations (Munn-Ponizovskii) with the results of Chevalley group representations (Curtis-Richen) we obtained the surprising result that irreducible modular representations of the monoid restrict to irreducible representations of the unit group. It is as if the finite group is somehow "dense" in the monoid, as in the geometric case. This led me, around 1998, to a complete classification of irreducible, modular representations of finite monoids of Lie type; along with an enumerative theory, relating these representations to the Weil zeta function of the adjoint quotient.

We mention here some related developments. Around 1990 Solomon began a study of the monoid Hecke-Iwahori algebra, initially for $M_n(\mathbb{F}_q)$. These algebras are semisimple, and they have very recently appeared (with Halverson et al.) in a solution of the Schur-Weyl duality theorem for quantum $gl_n(q)$.

Around 1990, S. Doty proved that the coordinate algebra of a reductive normal monoid M in characteristic $p > 0$ is a direct limit of generalized Schur algebras in the sense of Donkin. In particular, *Rep*(M) is a *highest weight category* in the sense of Cline, Parshall and Scott.

In 1994, É.B. Vinberg introduced some new ideas into the theory of algebraic monoids: abelianization, flat deformation, $Env(G_0)$ and the asymptotic semigroup $As(G_0)$. He also gave a new approach to the classification of reductive monoids.

Also around 1994, Rittatore in his Grenoble thesis systematically identified the entire theory of algebraic monoids as a part of the theory of spherical embeddings. He also extended much of Vinberg's work to characteristic $p > 0$, and later proved that any reductive, normal, algebraic monoid is Cohen-Macaulay.

In writing this survey I have tried to assess every contribution that impacts significantly on the theory of algebraic monoids. Hopefully, I have not improperly stated the work of any author. There is some difficulty on this point because there is a natural hierarchy of theories:

i) affine torus embeddings
ii) reductive algebraic monoids
iii) symmetric varieties
iv) spherical embeddings.

Indeed, this is obvious from the definitions (and a theorem of Vust, to get from iii) to iv)). One should also mention horospherical varieties along with this list. Each of these topics is a legitimate, well established discipline in its own right, with its own methods and techniques. Furthermore, many results about reductive monoids can be identified as the special case of some more general results about symmetric varieties or spherical embeddings. As we have already pointed out, this observation has led to some important work of Rittatore. He systematically identifies the theory of algebraic monoids as a special case within the theory of spherical embeddings. We describe his approach in § 5.3. We also identify the key ideas of embedding theory as they pertain to reductive monoids.

On the other hand, there are several features about algebraic monoids that have yet to be worked out for general spherical varieties:

i) The possible $G \times G$-orbits that could occur for some reductive monoid are easy to construct in explicit detail. See § 5.3.3 for some detail here. One can calculate the $B \times B$-orbits, and the adherence ordering on the set R of these orbits, in terms of the lattice of $G \times G$-orbits and the Bruhat ordering on the associated Weyl group, and certain of its subgroups.
ii) There is an abstract theory, due to Putcha, known as *monoids of Lie type* in the spirit of Tits' theory of BN-pairs.

These monoid constructions should ultimately work for more general spherical vartieties, when more is known about the "global" structure of spherical homogeneous spaces. It appears to be one of the wide open challenges to describe explicitly (in terms of the dense orbit) the possible spherical homogeneous spaces that could occur on the boundary of a given spherical variety.

Some of our results in Chapter 11, on the cell decomposition of the "wonderful" compactification X, have been obtained using other methods. Indeed,

Brion has obtained a cell decomposition of X using the method of Birula-Bialynicki.

This survey is organized as follows. Each of the next thirteen chapters is devoted to some particular theme directly related to algebraic monoids. Chapter 7, for example, is devoted to the problem of determining the orbit structure of reductive monoids. There is also a fifteenth chapter where we discuss several results that are directly related to the theory of algebraic monoids, but which require techniques beyond the scope of this survey.

There is no need to summarize every chapter in this introduction. We have already discussed the main results of the theory above. The reader should consult the table of contents for a description of each chapter and a guide to how the material is organized.

2

Background

In this chapter we assemble some of the major ideas and results from algebraic geometry, algebraic group theory and semigroup theory. This is intended to set the tone for the reader. It is intended also to provide some convenient references for the ensuing development. The theory of algebraic monoids is a rich blend of these three influences.

2.1 Algebraic Geometry

Algebraic monoids are affine, algebraic varieties with other structures attached to them. In this section, we introduce some basic concepts, such as varieties, morphisms, dimension and divisors. We assume in this section that K is an algebraically closed field.

2.1.1 Affine Varieties

We define *affine n-space* over K to be K^n, the set of all n tuples of elements of K. An element $P \in K^n$ is called a *point* and, if $P = (a_1, \dots, a_n)$, then a_i will be called the *coordinates* of P. Let $A = K[X_1, \dots, X_n]$ be the polynomial ring in n variables over K. We think of the elements of A as functions on K^n as follows: if $f(X_1, \dots, X_n) \in A$, then we define $f : K^n \to K$ by the rule $f(P) = f(a_1, \dots, a_n)$. Thus we can talk about the *zeros* of f, namely $Z(f) = \{P \in K^n | f(P) = 0\}$. If E is any subset of A, we define

$$Z(E) = \{P \in K^n | f(P) = 0 \text{ for all } f \in E\}.$$

Definition 2.1. *A subset X of K^n is called an* algebraic set *if $X = Z(E)$ for some subset E of A.*

Notice that, if $X = Z(E)$ is an algebraic set, then $X = Z(E_0)$ for some finite subset E_0 of E. Indeed, A is a Noetherian ring, and thus

$Z(E) = Z((E))$, where (E) denotes the ideal generated by E. But then $(E) = (f_1, \ldots, f_n)$ is finitely generated by the Noetherian condition and thus $Z((E)) = Z(\{f_1, \ldots, f_m\})$.

Proposition 2.2. *The union of two algebraic sets is algebraic. The intersection of any collection of algebraic sets is algebraic. The empty set is algebraic. The whole space is algebraic.*

Proof. If $X = Z(E)$ and $Y = Z(F)$, then $X \cup Y = Z(EF)$, where $EF = \{fg \,|f \in E \text{ and } g \in F\}$. If $X_\alpha = Z(E_\alpha)$, then $\cap X_\alpha = Z(\cup E_\alpha)$. $\phi = Z(1)$ and $K^n = Z(0)$.

Definition 2.3. *The* Zariski topology *on K^n is the topology on K^n defined by taking as open sets the complements of algebraic sets. By Proposition 2.2 this is a topology on K^n.*

Example 2.4. Consider the Zariski topology on K. In this case, $A = K[X]$, and it is well known that every ideal of A is principal. Thus every algebraic set Z is the zero locus of a single polynomial $f \in A$. Furthermore, since K is algebraically closed, f factors as $f(X) = c(X - a_1)\ldots(X - a_n)$ with $c, a_1, \ldots, a_n \in K$. Hence $Z = \{a_1, \ldots, a_n\}$. Thus the Zariski topology on K is the *cofinite* topology.

Definition 2.5. *A nonempty subset of a topological space X is called* irreducible *if it cannot be expressed as the union $X = X_1 \cup X_2$ of two, nonempty, proper closed subsets of X.*

Example 2.6. K is irreducible because any proper closed subset of K is finite, while K is algebraically closed, and therefore infinite.

Theorem 2.7. (Hilbert's Vanishing Theorem) *Let K be an algebraically closed field, let $\mathfrak{a}$ be an ideal of $A = K[X_1, \ldots, X_n]$, and let $f \in A$ be a polynomial which vanishes at all points of $Z(\mathfrak{a})$. Then $f^r \in \mathfrak{a}$ for some integer $r > 0$.*

Proof. See Atiyah-Macdonald [2] page 85.

Thus, there is an inclusion-reversing correspondence between algebraic sets in K^n and radical ideals of $A = K[X_1, \ldots, X_n]$. It is easy to check that, under this correspondence, prime ideals correspond to irreducible closed subsets.

Example 2.8. K^n is irreducible, since it corresponds to the zero ideal in A.

Example 2.9. If f is an irreducible polynomial in $A = K[X_1, \ldots, X_n]$, then $Z(f)$ is an irreducible, algebraic subset of K^n of codimension one. $Z(f)$ is called a *hypersurface.*

If $Y \subseteq K^n$ we define the **ideal** of Y by

$$I(Y) = \{f \in A | f(P) = 0 \text{ for all } P \in Y\}.$$

Definition 2.10. *If Y is an affine, algebraic set, the* affine coordinate ring *of Y is $K[Y] = A/I(Y)$.*

We now study the Zariski topology on affine varieties.

Definition 2.11. *A topological space X is called* noetherian *if it satisfies the descending chain condition on closed sets: for any sequence $Y_1 \supseteq Y_2 \supseteq \dots$ of closed sets, there exists an integer $r > 0$ such that $Y_r = Y_{r+1} = \dots$.*

It is easily checked that any affine algebraic set Y is a noetherian topological space. Indeed, this follows directly from the fact that $K[X_1, \dots, X_n]$ is a **noetherian ring** which, by definition, is a ring which satisfies the **ascending chain condition** on ideals. Any descending chain of closed subsets of Y determines an ascending chain of ideals of $K[Y]$.

Noetherian topological spaces behave differently from Hausdorff topological spaces.

Proposition 2.12. *Let X be a noetherian topological space and let $Y \subseteq X$ be a closed subset of X. Then Y can be expressed as a finite union $Y = Y_1 \cup \dots \cup Y_r$ of irreducible subsets. If we insist that $Y_i \not\subseteq Y_j$ for $i \neq j$, then the Y_i are uniquely determined.*

The Y_i are called the *irreducible components* of Y.

Proof. To prove existence of such a decomposition of Y one uses *Zorn's lemma.* Let $\mathcal{S}$ be the set of nonempty closed subsets of X which cannot be written as a finite union of irreducible, closed subsets. If $\mathcal{S}$ is nonempty, it must have a minimal element, since X is a noetherian topological space. Let Y be such a minimal element. Then Y must be reducible, and therefore we can write $Y = U \cup V$ where U and V are proper closed subsets of Y. By minimality of Y, each of U and V can be written as a finite union of irreducible closed subsets, and hence Y also: a contradiction. This establishes the first part of the claim. We leave the rest of the proof to the reader.

Corollary 2.13. *Any algebraic set in K^n can be expressed uniquely as a union of irreducible closed subsets, no one containing the other.*

2.1.2 Dimension Theory

We begin with a definition.

Definition 2.14. *a) Let X be a topological space. The* dimension *of X is the supremum of all integers n such that there exists a chain $Z_0 \subseteq Z_1 \subseteq \dots \subseteq Z_n$ of distinct, irreducible closed subsets of X. We define the dimension of an affine variety to be its dimension in this sense.*

b) In a commutative ring A, the height *of a prime ideal $\mathfrak{p}$ is the supremum of all integers n such that there is a chain $\mathfrak{p}_0 \subset \mathfrak{p}_1 \subset \dots \subset \mathfrak{p}_n = \mathfrak{p}$ of distinct prime ideals. The* dimension *or* Krull dimension *of A is the supremum of the heights of all prime ideals of A.*

Proposition 2.15. *If X is an affine algebraic set, then the dimension of X is equal to the dimension of its affine coordinate ring $K[X]$.*

Proof. There is a one-to-one inclusion-reversing correspondence between the prime ideals of $K[X]$ and the irreducible closed subsets of X.

Remark 2.16. a) It follows from Chapter 11 of [2] that the dimension of $K[X]$ is equal to the transcendence degree of the fraction field $K(X)$ of $K[X]$ over K.
b) It follows from a) above that the dimension of K^n is n.
c) If A is a noetherian ring and $f \in A$ is a regular element, then $\dim(A/(f)) = \dim(A) - 1$. See page 122 of [2].

2.1.3 Divisor Class Groups

The class group ultimately contains a subtle mixture of local and global information about a normal, algebraic variety. In general, it is not easy to calculate these class groups. But on the other hand, it is often possible to compute the class group of a variety which can be expressed as the union of well-behaved subvarieties.

Our general reference for this section is Fossum's monograph [29]. Also, Section 6 of Chapter II of [38] is a good introduction from a more geometric point of view.

A commutative ring A is called an **integral domain** if, for any $x, y \in A\backslash\{0\}$, $xy \neq 0$. It is easy to check that A is an integral domain if and only if the zero ideal of A is a prime ideal. Let A be a noetherian integral domain. We say that A is **normal** if it is integrally closed in its field $K(A)$ of fractions. We call an irreducible, algebraic variety X **normal** if its coordinate ring $A = K[X]$ is a normal integral domain. A minimal, nonzero prime ideal $\mathfrak{p}$ of A is called a **height one** prime ideal. If X is an algebraic variety over K then the height one primes of A are in one-to-one correspondence with the closed irreducible subvarieties Y of X of codimension one. We call these subvarieties **prime divisors**. It follows from Theorem 38, page 124 of [57], that

$$A = \bigcap_{ht(\mathfrak{p})=1} A_{\mathfrak{p}}.$$

Furthermore, each $A_{\mathfrak{p}}$ is a discrete valuation ring of A. We denote by

$$\nu_Y : K(A)^* \to \mathbb{Z}$$

the discrete valuation on $K(A)$ determined by $\mathfrak{p}$ and $Y = Spec(A/\mathfrak{p})$. If $f \in K(A)$, it is easy to check that

$$\nu_Y(f) = 0$$

for all but a finite number of prime divisors Y of X.

Definition 2.17. *If X is a normal, algebraic variety, let $Div(X)$ be the free abelian group with basis $\{Y \mid Y$ is a prime divisor of $X\}$. If $f \in K(X)$ we define the* divisor *of f, denoted $div(f)$, by*

$$div(f) = \sum \nu_Y(f)Y,$$

where the sum is taken over all prime divisors of X. We refer to $div(f)$ as a principal divisor, *and denote by $Prin(X) \subseteq Div(X)$ the subgroup of principal divisors. Finally, we define the* divisor class group *of X:*

$$Cl(X) = Div(X)/Prin(X).$$

Notice that we have an exact sequence of abelian groups:

$$0 \to K^* \to K(X)^* \to Prin(X) \to Div(X) \to Cl(X) \to 0.$$

Example 2.18. Let X be a normal, irreducible, algebraic variety. Then the following are equivalent:

a) $Cl(X) = 0$.
b) Every height one prime $\mathfrak{p}$ of $K[X]$ is principal.
c) $K[X]$ is a unique factorization domain.

In particular, K^n has trivial divisor class group.

Proposition 2.19. *Let X be irreducible and normal, and let Z be a proper, closed subvariety of X. Let $U = X \backslash Z$.*

a) There is a surjective morphism $Cl(X) \to Cl(U)$ defined by $Y \to Y \cap U$ if $Y \cap U$ is nonempty, and zero otherwise.
b) If $codim_X(Z) \geq 2$, then $Cl(X) \to Cl(U)$ is an isomorphism.
c) If $Z = \cup_i Z_i$ is a union of prime divisors, then there is an exact sequence

$$\oplus_i \mathbb{Z} \to Cl(X) \to Cl(U) \to 0$$

where the first map is defined by $(a_1, \ldots, a_n) \to \sum a_i Z_i$. In particular, if $Cl(U)$ is trivial, then $Cl(X)$ is generated by $\{Z_i\}$.

Proof. For a) notice that every prime divisor of U is the restriction of its closure in X. The result in b) follows since $Prin(U) = Prin(X)$ and $Div(U) = Div(X)$. For c), notice that the kernel of $Div(X) \to Div(U)$ is generated by $\{Z_i\}$.

Example 2.20. If $X = \mathbb{P}^n$, then $Cl(X) = \mathbb{Z}$. Indeed, let $H \subseteq X$ be a linear hypersurface. Then by c) above, $Cl(X)$ is generated by the class of H, since $X \backslash H = K^n$. On the other hand, each divisor Y of X has a well defined degree determined by the degree of its defining equation. But any rational function on X has degree zero, being the the quotient of two homogeneous polynomials of the same degree. Hence $degree : Cl(X) \to \mathbb{Z}$ is an isomorphism.

2.1.4 Morphisms

In this section we acquaint the reader with some of the basic facts about morphisms of algebraic varieties. Our discussion is mainly concerned with affine varieties and affine morphisms. This simplifies the discussion significantly.

Definition 2.21. *a) Let X be an affine variety with coordinate ring $K[X]$. A function $f : X \to K$ is* regular at a point *$P \in X$ if there is an open subset $U \subseteq X$ with $P \in U$, and $g, h \in K[X]$ such that $f = g/h$ on U.*
b) We say that f is regular on *X if it is regular at every point of X.*

Definition 2.22. *Let X and Y be irreducible affine varieties. A* morphism *$\psi : X \to Y$ is a continuous function such that, for every open subset $U \subseteq Y$ and every regular function $f : U \to K$, $f \circ \psi : \psi^{-1}(U) \to K$ is a regular function.*

Proposition 2.23. *Let X and Y be affine algebraic varieties with coordinate rings $K[X]$ and $K[Y]$ respectively. Define*

$$\gamma : Hom(X, Y) \to Hom(K[Y], K[X])$$

by $\gamma(f) = f^*$, where $f^*(h) = h \circ f$. Then γ is an bijection. Here Hom on the left means morphisms of varieties, and Hom on the right means morphisms of K-algebras.

Proof. We give a sketch. See page 19 of [38] for more details. The map γ is well defined since $K[X]$ is canonically identified with the ring of regular functions on X. Furthermore, γ is clearly one-to-one.

Conversely, given a homomorphism $\psi : K[Y] \to K[X]$ of K-algebras, define $\psi^* : X \to Y$ as follows. For $x \in X$, define ϵ_x by $\epsilon_x(g) = g(x)$. Then define $\psi^*(x) = \epsilon_x \circ \psi$. One then checks that ψ^* is a morphism, and that $\gamma(\psi^*) = \psi$.

A version of the above result is true even if X is not affine. In that case, let $\mathcal{O}(X)$ be the ring of regular functions on X. Then

$$\gamma : Hom(X, Y) \to Hom(K[Y], \mathcal{O}(X))$$

is a bijection. See Proposition 3.5 of Chapter I of [38] for more details.

We now distinguish certain classes of morphisms that will be important in our later discussions.

Definition 2.24. *a) A morphism $f : X \to Y$ is* finite *if $f^* : K[Y] \to K[X]$ makes $K[X]$ into a finitely generated module over $K[Y]$.*
b) A morphism $f : X \to Y$ is dominant *if $f(X) \subseteq Y$ is a dense subset. Notice that this is equivalent to saying that $f^* : K[Y] \to K[X]$ is injective.*
c) A dominant morphism $f : X \to Y$, between irreducible varieties, is birational *if f induces an isomorphism $f^* : K(Y) \to K(X)$ of function fields.*

d) *A morphism $f : X \to Y$ is* flat *if the functor $F(M) = M \otimes_{K[Y]} K[X]$, from $K[X]$-modules to $K[Y]$-modules, is exact.*

Remark 2.25. a) A finite morphism has finite fibres.

b) A finite dominant morphism $f : X \to Y$ induces a K-algebra homomorphism $f^* : K(Y) \to K(X)$ of function fields. The typical fibre has s points in it, where s is the separable degree of f.

c) If $f : X \to Y$ is a birational morphism, then there are open subsets U of X and V of Y such that $f|U : U \to V$ is an isomorphism.

d) Let X and Y be affine varieties with graded coordinate algebras $K[X] = \sum_{n \geq 0} A_n$ and $K[Y] = \sum_{n \geq 0} B_n$, respectively. Assume also that $A_0 = B_0 = K$. Then each of X and Y has a cone point $0_X \in X$ and $0_Y \in Y$. Let $f : X \to Y$ be a morphism of varieties such that f^* is a homomorphism of graded K-algebras. Then f is a finite morphism if and only if $f^{-1}(0_Y) = 0_X$.

e) A flat surjective morphism is open, and has equidimensional fibres.

Given a normal, irreducible, affine variety X, it is sometimes possible to construct a morphism $f : U \to Y$ from some open subset U of X to the affine variety Y. On the other hand, we would then like to know whether f extends to a morphism $\overline{f} : X \to Y$, without actually constructing this extension. The following **codimension two condition** gives us a very useful criterion.

Theorem 2.26. *Let X be a normal, irreducible, affine variety, and assume that $U \subseteq X$ is an open subset such that $codim_X(X \backslash U) \geq 2$. If $f : U \to Y$ is a morphism to the affine variety Y, then f extends uniquely to a morphism $\overline{f} : X \to Y$.*

Proof. Our assumptions give us a K algebra homomorphism $f^* : K[Y] \to \mathcal{O}(U)$. However, by a previous remark in this section

$$A = \bigcap_{ht(\mathfrak{p})=1} A_{\mathfrak{p}},$$

where $A = K[X]$. But $\mathcal{O}(U) = \bigcap_{ht(\mathfrak{p})=1} A_{\mathfrak{p}}$, since $codim_X(X \backslash U) \geq 2$. Thus $K[X] = \mathcal{O}(U)$.

Example 2.27. Let $Z = K^2$, and let $U = Z \backslash \{0\}$. Then it is easy to check that $\mathcal{O}(U) = K[X, Y, X^{-1}] \cap K[X, Y, Y^{-1}] = K[X, Y] = \mathcal{O}(Z)$.

The codimension 2 condition has been used very effectively by Grosshans [34] in his work on invariant theory.

It is useful in characteristic $p > 0$ to keep track of the separable degree of a morphism. Our definition of separable is not the most general one, but it is good enough for our purposes.

Definition 2.28. *Let $f : X \to Y$ be a dominant morphism of irreducible, algebraic varieties. Assume that f is generically finite. This means that $f^* : K(Y) \to K(X)$ is a finite extension of fields. We say that f is* separable *if f^* is a separable extension of fields.*

It turns out that, if $f : X \to Y$ is generically finite, then there is an open subset U of Y such that $|f^{-1}(y)|$ is constant for $y \in U$. If further f is separable then the degree of $K(X)$ over $K(Y)$ is this common value.

Notice in particular that, if f is injective, dominant and separable, then it is birational. Also notice that any generically finite morphism is separable in characteristic zero.

Theorem 2.29. (Zariski's Main Theorem) *Let $f : X \to Y$ be a birational morphism between irreducible varieties. Assume that f is finite-to-one and that Y is normal. Then f is an open embedding. In particular, if f is also surjective, then it is an isomorphism of varieties.*

For a development of this Theorem see Corollary 11.4, Chapter III of [38].

2.2 Algebraic Groups

In this section we (re)acquaint the reader with the fundamentals of algebraic group theory. The reader who is unfamiliar with algebraic groups and their finite dimensional representations should consult [7, 40, 69, 134]. Algebraic group theory is the "generic point" of any theory of algebraic monoids.

Obviously, we cannot state or prove everything we need here. So we try to assemble the main constructions and results that are particularly relevant to the development of the theory of algebraic monoids. Notice, in particular, that we are interested only in *affine* algebraic groups.

As usual we assume that our algebraic varieties are defined over the algebraically closed field K.

2.2.1 Algebraic Groups

Definition 2.30. *Let G be an algebraic variety. Assume that we have morphisms of algebraic varieties $m : G \times G \to G$, $m(x, y) = xy$, and $i : G \to G$, $i(x) = x^{-1}$, such that G is a group with m as multiplication and i as inverse. Then (G, m, i) is called an* **algebraic group**.

Remark 2.31. Let G be an algebraic group.

a) There are obvious notions of morphism and isomorphism of algebraic groups.
b) Any algebraic group is a smooth variety.
c) The direct product of algebraic groups is an algebraic group.

d) Any closed subgroup H of G is an algebraic group with the group structure it inherits from G.
e) If $\rho : G \to H$ is a morphism of algebraic groups, then the kernel K and image N of ρ are algebraic groups. Furthermore, $\dim(G)=\dim(K)+\dim(N)$.
f) If N is a closed, normal subgroup of G, then G/N has the unique structure of an algebraic group such that the canonical morphism $\pi : G \to G/N$ is a morphism of algebraic groups.
g) The irreducible components of G are in fact the connected components. So there is a unique, connected component of the identity, denoted G^0. G^0 is normal in G and has finite index in G.

Example 2.32. a) K^*, the multiplicative group of nonzero elements of K.
b) $(K, +)$, the *additive group*.
c) $T_n(K)$, the group of upper-triangular invertible $n \times n$ matrices.
d) $D_n(K)$, the group of diagonal invertible $n \times n$ matrices.
e) $U_n(K)$, the group of unipotent upper-triangular $n \times n$ matrices.
f) $Gl_n(K)$, the group of $n \times n$ invertible matrices.

Any algebraic group has certain distinguished subgroups associated with it, suggested already by the above examples. We first define these different types of groups.

Definition 2.33. *Let G be a connected, algebraic group.*

a) G is solvable *if it is solvable as a group.*
b) G is a D-group *or a* torus *if its coordinate algebra is generated by characters. A* character *is a morphism $\chi : G \to K^*$.*
c) G is nilpotent *if it is nilpotent as a group.*
d) G is unipotent *if, for any morphism $\rho : G \to Gl_n(K)$, there is a nonzero vector $v \in K^n$ such that $\rho(g)(v) = v$ for any $g \in G$. Any unipotent algebraic group is nilpotent.*

$T_n(K)$ is solvable. By the Lie-Kolchin Theorem [40], any connected, solvable group is isomorphic to a closed subgoup of $T_n(K)$ for some n.

$D_n(K)$ is a D-group. Any D-group is isomorphic to a closed subgroup of $D_n(K)$ for some n.

$U_n(K)$ is unipotent. Any unipotent algebraic group is isomorphic to a closed subgroup of $U_n(K)$ for some n.

Each of the groups mentioned above is a maximal subgroup, of the given type, of $Gl_n(K)$.

Definition 2.34. (The Radical) *Let G be a connected algebraic group. G has a maximal, connected, unipotent, normal subgroup, denoted $R_u(G)$. $R_u(G)$ is called the* unipotent radical *of G. G has a maximal, connected, solvable normal subgroup, denoted $R(G)$. $R(G)$ is called the* radical *of G.*

In each case, factoring out the radical yields a group with trivial radical of that type. A group G is called **reductive** if the unipotent radical is trivial, and **semisimple** if the radical is trivial. Reductive groups are the most important class of algebraic groups.

Any algebraic group has maximal, connected, solvable (or unipotent or diagonalizable) subgroups. This would be a minor issue if there was no way to compare any two of these maximal subgroups. However, we have the following extremely useful conjugacy thereom. This allows one to associate with each algebraic group, exactly one set of structure constants for each type of subgroup. This, ultimately leads to a classification of semisimple algebraic groups.

Theorem 2.35. (Conjugacy Theorems) *Let G be a connected algebraic group, and let H and K be two maximal, connected, solvable (or diagonalizable, or unipotent) subgroups of G. Then there exists $g \in G$ such that $gHg^{-1} = K$. Each maximal, connected, solvable subgroup B is the semidirect product of its unipotent radical and any of its maximal tori. The maximal tori of B continue to be maximal tori of G. The unipotent radical of B is a maximal, unipotent subgroup of G.*

Proof. See Theorem 21.3 and Corollary 21.3A of [40] .

The maximal solvable connected subgroups are called **Borel subgroups**. Any solvable, connected group G is isomorphic to the semidirect product $G = TU$ of its unipotent radical U and any of its maximal tori T.

One method of proof of the conjugacy of Borel subgroups is the Borel Fixed Point Theorem.

Theorem 2.36. (Borel Fixed Point Theorem). *Let G be a solvable, connected algebraic group acting on the complete variety X. Then G has a fixed point.*

Proof. Let $H = (G, G)$. Since G is solvable, $\dim(H) < \dim(G)$. Hence by induction on the dimension of G, H has a fixed point on X. If we let Y be the set of fixed points of H on X, then the commutative algebraic group G/H acts on the nonempty complete variety Y. We are thereby reduced to the case of a commutative group $A = G/H$. But now the action of A on Y has orbits of minimal dimension, which are closed and irreducible. On the other hand, these orbits are affine. But any irreducible, complete, affine variety is a point.

Corollary 2.37. *Let B and B' be two Borel subgroups of the algebraic group G. Then there exists $g \in G$ such that $gBg^{-1} = B'$.*

Proof. Let B' be a Borel subgroup of G of maximal dimension, and let B be any other Borel subgroup of G. Consider the action $B' \times G/B \to G/B$ defined by $(b, gB) \to bgG$. By the Borel Fixed Point Theorem, B' has a fixed point gB on G/B, since G/B is a projective variety. So $B'gB = gB$, and thus $B'gBg^{-1} = gBg^{-1}$. Hence, $B' \subseteq gBg^{-1}$ giving $B' = gBg^{-1}$.

2.2.2 Root Systems, Weyl Groups and Dynkin Diagrams

There is a much-studied classification of semisimple groups that depends on discrete data obtained from the maximal torus and how it acts on the maximal unipotent subgroup of its ambient Borel subgroup. This classification inevitably involves root sytems, Weyl groups and Dynkin diagrams. The reader is advised to acquire familiarity with at least one of the many textbooks on this much celebrated theory. Reference [69] contains many specific facts that are useful in classification problems related to algebraic monoids, and [40] develops the theory in detail from a modest background in linear algebra and algebraic geometry. Our summary here is brief, and is intended only for convenient, quick reference. In particular, very little is said about Lie algebras. For more details the reader should consult [7, 40, 69].

The list of simple, algebraic groups is amazingly short, and does not depend on the (algebraically closed) field K. In fact, each group can be defined over $\mathbb{Z}$ in such a way that it will specialize to yield the correct (split) group over any ring. There are four infinite families of simple groups, and five exceptional groups. Each group has a diagram associated with it, known as its **Dynkin diagram**. The Dynkin diagram efficiently codes the structural information needed to construct the group.

Let G be a semisimple, algebraic group. Let B be a Borel subgroup with maximal torus $T \subseteq B$ and unipotent radical U. T acts on U by inner automorphisms, $u \to tut^{-1}$. This action induces an action of T on the tangent space $\mathfrak{u}$ of U. Since T is a D-group, $\mathfrak{u}$ decomposes into weight spaces indexed by certain characters $\Phi^+ \subseteq X(T)$, known as (positive) **roots**:

$$\mathfrak{u} = \oplus_{\alpha \in \Phi^+} \mathfrak{g}_\alpha$$

We let $\Phi = \Phi^+ \cup -\Phi^+$.

Theorem 2.38. *a) dim($\mathfrak{g}_\alpha$)=1, for each $\alpha \in \Phi^+$.*
b) There is a unique, closed T-stable subgroup U_α of U whose tangent space at the identity of U is $\mathfrak{g}_\alpha$.
c) There is a unique, Borel subgroup B^-, called the Borel subgroup opposite *to B (relative to T), such that $T \subseteq B^-$ and $B \cap B^- = T$.*
d) If U^- is the unipotent radical of B^-, the set of weights of T on $\mathfrak{u}^-$ is $-\Phi^+$.
e) G is generated as a group by the groups $U_\alpha, \alpha \in \Phi$ and T.
e) Φ generates a subgroup of finite index in $X(T)$.

Example 2.39. Let $G = Sl_n(K)$, and let $B = T_n(K) \cap G$, $U = U_n(K)$ and $T = D_n(K) \cap G$. Then $B^- = LT_n(K) \cap G$, where $LT_n(K)$ is the group of invertible lower-triangular matrices. One checks easily that $\Phi^+ = \{\alpha_{i,j} | i > j\}$ and $\Phi^- = \{\alpha_{i,j} | i < j\}$. Here, $\alpha_{i,j}(t_1, \ldots, t_n) = t_i t_j^{-1}$ and $U_{i,j} = \{I_n + aE_{i,j} | a \in K\}$, where $E_{i,j}$ is the elementary matrix with one non zero entry in the (i,j)-position.

The above theorem ultimately leads to the following definition of a root system. For convenience, these objects are usually defined over $\mathbb{R}$.

Definition 2.40. *A* root system *is a real vector space E together with a finite subset Φ, called* roots, *satisfying:*

a) Φ spans E, and does not contain zero.
b) If $\alpha \in \Phi$, the only other multiple of α in Φ is $-\alpha$.
c) If $\alpha \in \Phi$, there is a reflection $\sigma_\alpha : E \to E$ such that $\sigma_\alpha(\alpha) = -\alpha$, and σ_α leaves Φ stable.
d) If $\alpha, \beta \in \Phi$, then $\sigma_\alpha(\beta) - \beta$ is an integral multiple of α.

Remark 2.41. a) The group W generated by $\{\sigma_\alpha | \alpha \in \Phi\}$ is called the **Weyl group**.
b) A subset $\Delta = \{\alpha_1, \ldots, \alpha_r\}$ is called a **base** if Δ is a basis of E, and each $\alpha \in \Phi$ has a unique expression of the form $\alpha = \sum c_i \alpha_i$, where the c_i are integers, either all nonnegative or all nonpositive. Bases exist, every root is in at least one base, and W permutes them simply transitively.
c) The elements of Δ are called **simple roots**, and the corresponding reflectioons are called **simple reflections**.
d) W is already generated by $\{\sigma_\alpha | \alpha \in \Delta\}$, and as such it is a **Coxeter group**.
e) There is an inner product (α, β) on E relative to which W is a group of orthogonal transformations. For σ_α we obtain $\sigma_\alpha(\beta) = \beta - <\beta, \alpha> \alpha$, where $<\beta, \alpha>= 2(\beta, \alpha)/(\alpha, \alpha)$.
f) If G is a semisimple group with maximal torus T, let $E = X(T) \otimes \mathbb{R}$. Then (E, Φ), as in the above theorem, is a root system. The Weyl group of this root system is canonically isomorphic to $N_G(T)/T$. G is generated, as a group, by B and $N_G(T)$.
g) Φ is called **irreducible** if it cannot be partitioned into a union of two, mutually orthogonal, proper subsets.

Up to isomorphism, the irreducible root systems correspond to the Dynkin diagrams, which are depicted in Figure 2.1. Each irreducible root system corresponds to a simple algebraic group.

The numbered nodes (circles) in each diagram correspond to the simple roots. Nodes corresponding to α and β are joined by $<\alpha, \beta><\beta, \alpha>$ bonds. There is an arrow pointing to the shorter of the two roots, if indeed the roots are of different length. Notice that α and β can be joined by 0, 1, 2, or 3 bonds, according to whether the order of $\sigma_\alpha \sigma_\beta \in W$ is 2, 3, 4, or 6.

For convenience and completeness, we have depicted the **extended** Dynkin diagrams. The extra node (circle with a "$\times$") corresponds to the highest root, which is also the highest weight of the adjoint representation.

It is easy to see that the information embodied in the Dynkin diagram is equivalent to the information embodied in the **Cartan matrix**:

$$<\alpha, \beta>; \ \alpha, \beta \in \Delta.$$

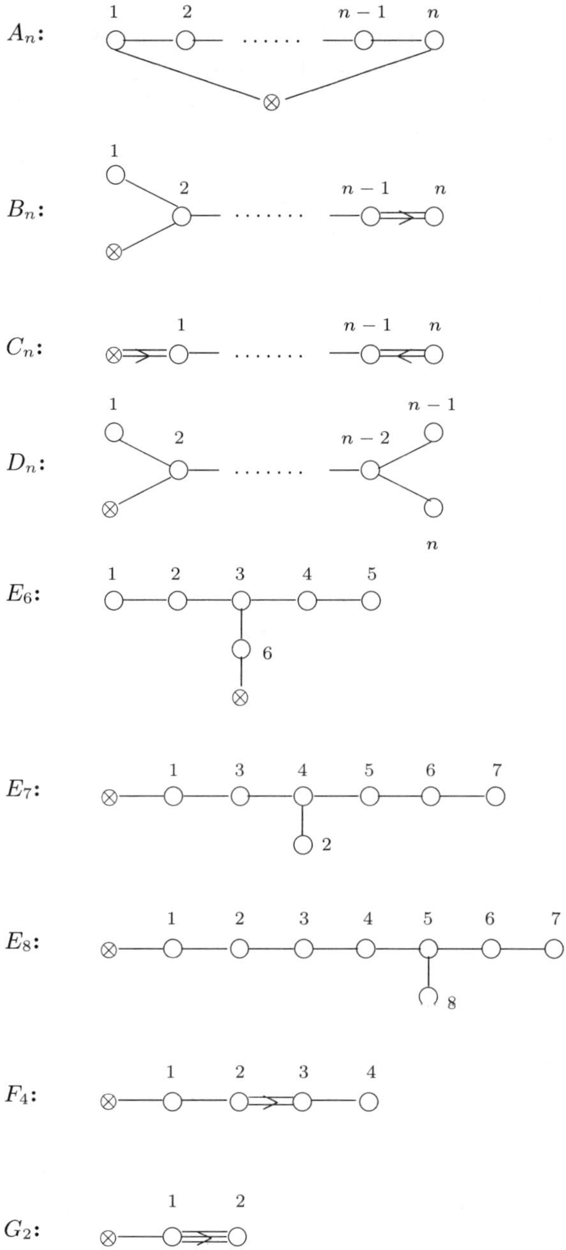

Fig. 2.1. The extended Dynkin diagrams of the simple algebraic groups.

The set of **fundamental dominant weights** $\{\lambda_1, \ldots, \lambda_r\}$ is defined so that $< \lambda_i, \alpha_j >= \delta_{i,j}$ (Kronecker delta). The **dominant weights** are the $\mathbb{Z}$-linear combinations $\lambda = \sum c_i \lambda_i$ with each $c_i \geq 0$. Each weight is conjugate under W to exactly one dominant weight. $X(T)$ has finite index in the set of weights. The Cartan matrix is the coefficient matrix for expressing the fundamental dominant weights in terms of the simple roots. It can also be used to define a presentation of $\mathfrak{g}$ via generators and relators.

2.2.3 Tits System and Bruhat Decomposition

Inspired by work of Chevalley [14], Tits [140] devised an efficient set of axioms to describe the structure of the simple Chevalley groups and other simple algebraic groups. The resulting theory, known as **Tits systems** or BN-pairs, is extremely efficient and far-reaching. It is essential in the development of Putcha's theory of monoids of Lie type.

Definition 2.42. (Tits System) *Let G be a group generated by two subgroups B and N, where $T = B \cap N$ is a normal subgroup of N. Let $W = N/T$, and assume that $S \subseteq W$ is a set of elements of order two of W. By standard abuse of language, we write wB for $w \in W$. This is allowed since two representatives of w in N differ by an element of T, which is contained in B. We say that (G, B, N, S) is a* Tits system *if*

a) for $s \in S$ and $w \in W$, $sBw \subset BwB \cup BswB$;
b) for $s \in S$, $sBs \neq B$.

W is the Weyl group *of the system, and $|S|$ is the* rank. *A subgroup of G conjugate to B is called a* Borel subgroup *of G.*

Example 2.43. Let G be a reductive group, and B a Borel subgroup of G containing the maximal torus T. Let $N = N_G(T)$ and let S be the set of simple reflections corresponding to the base Δ determined by T and B. Then (G, B, N, S) is a Tits system.

Theorem 2.44. *Let (G, B, N, S) be a Tits system with Weyl group W. For $I \subseteq S$ let W_I be the subgroup of W generated by I, and let $P_I = BW_IB$.*

a) P_I is a subgroup of G. In particular, $P_S = G$.
b) For $v, w \in W$, $BvB = BwB$ if and only if $v = w$. In particular, $sBw \subset BswB$ if and only if $sBw \cap BwB = \phi$.

The subgroups P_I are called **parabolic** subgroups of G.

Definition 2.45. *For $w \in W$, define the* length of w *relative to S as $l(w) = min\{k | w = s_1 \ldots s_k,\ s_i \in S\}$.*

Theorem 2.46. *a) The only subgroups of G containing B are the P_I.*
c) If P_I is conjugate to P_J, then $I = J$.

c) The following are equivalent.
i) $I = J$.
ii) $W_I = W_J$.
iii) $P_I = P_J$.
d) $N_G(P) = P$.

2.2.4 Representations

In this section, we describe the set of irreducible, rational representations of a semisimple group G. The case of a reductive group is only slightly more complicated. As usual, we let T be a maximal torus of G and $B = TU$ a Borel subgoup of G containing T. Let $B^- = TU^-$ be the opposite Borel subgroup containing T, and Δ the the base of Φ determined by B. Let

$$\rho : G \to Gl(V)$$

be a rational representation of G. The **weights** of ρ are the characters of T associated with the eigenspaces of the action of T on V. Then $V = \oplus V_\lambda$, where $V_\lambda = \{v \in V | \rho(t)(v) = \lambda(t)v, t \in T\}$. By the Lie-Kolchin theorem, there is a one-dimensional subspace L of V such that $\rho(B)(L) = L$. Then L is pointwise fixed by the unipotent radical of B. A nonzero vector v in L is called a **highest weight vector**.

Proposition 2.47. *Let V be a nonzero, rational G-module, and let v be a highest weight vector. Let V' be the submodule of V generated by v. Then the weights of V' are of the form $\lambda - \sum c_\alpha \alpha$ where $\alpha \in \Delta$ and the c_α are non-negative integers. Furthermore, dim(V'_λ)=1, and V' has a unique, maximal, proper submodule M. Consequently, V'/M is an irreducible G-module.*

Proof. Since $\rho(U)(v) = \{v\}$, V' is spanned by $\rho(U^-)(v)$. But U^- is a product of U_α's with $\alpha \in -\Phi^+$, and so applying U^- to a vector of weight λ results in a vector of the form $v+u$. But the components of u have weight $\lambda - \sum c_\alpha \alpha \neq \lambda$, where $\alpha \in -\Phi^+$ and the c_α are non negative. In particular, dim(V'_λ)=1.

Any proper submodule of V'_λ cannot contain v, and consequently it cannot contain any vectors of weight λ. So take M to be the sum of all proper submodules of V'.

The weight λ is called the **highest weight** of V', and V' is called a **highest weight module**. The above proposition shows that if we order the weights of V' as follows:

$$\lambda > \mu$$

whenever $\lambda - \mu$ is a sum of positive roots, then λ is greater than all other weights of V' for this partial ordering. It turns out that this weight λ is actually a dominant weight.

Theorem 2.48. *a) Let V be an irreducible, rational G-module. There is a unique B-stable one-dimensional subspace spanned by a highest weight vector v with dominant weight λ. All other weights of V are of the form $\lambda - \gamma$, where γ is a sum of positive roots.*

b) If V' is another irreducible rational G-module with highest weight λ', then V and V' are isomorphic if and only if $\lambda = \lambda'$.

c) Let $\lambda \in X(T)$ be a dominant weight. Then there exists an irreducible G-module V_λ of highest weight λ.

Proof. For a), we already have everything but the uniqueness. But there cannot be two different, highest weights. For b), if V and V' are two irreducible G-modules with highest weight λ, let $v \in V$ and $v' \in V'$ be the respective highest weight vectors. It is easy to construct a highest weight module V'' inside $V \oplus V'$ that projects onto both V and V'. But it has a unique maximal submodule, $M \subset V''$. Then both V and V' are isomorphic to V''/M.

To prove c), define

$$H^0(\lambda) = \{f \in K[G] | f(xy) = \lambda(x)f(y) \text{ for } x \in B^-,\ y \in G\}.$$

One checks that $H^0(\lambda)$ is a subspace of $K[G]$ stable under right translation. It is possible to find a function $f \in H^0(\lambda)$ such that $f(xy) = \lambda(y)f(x)$ for all $x \in G$ and $y \in B$. Here, we may think of λ as a character on B by declaring $\lambda(u) = 1$ for $u \in U$. It turns out that the submodule of $H^0(\lambda)$ generated by this f is the sought after irreducible representation. In characteristic zero, $H^0(\lambda)$ is actually irreducible.

Remark 2.49. **(Borel-Weil-Bott Theory)** The entity $H^0(\lambda)$ in the proof of part c) above can be interpreted geometrically. If λ is interpreted as above, as a character $\lambda : B \to K^*$, we can define a line bundle on G/B as follows. Let B act on $G \times K^*$ by the rule $b * (g,t) = (gb^{-1}, \lambda(b)t)$. Let $L(\lambda) = \{[g,t] | g \in G,\ t \in T\}$ be the quotient space of this action. We then have a canonical projection $\pi : L(\lambda) \to G/B$ defined by setting $\pi([g,t]) = gB$. Then π is a principal $\mathbb{G}_m$-bundle over G/B. We let $\mathfrak{L}(\lambda)$ be the sheaf on G/B associated with π. Notice that λ is not required to be dominant for this construction. The Borel-Weil Theorem states that:

a) $H^0(\lambda)$ is the space of sheaf-theoretic global sections of $\mathfrak{L}(\lambda)$.
b) $H^0(\lambda)$ is nonzero if and only if λ is dominant; and irreducible if char(K)=0.
c) The correspondence $\lambda \to \mathfrak{L}(\lambda)$ determines a one-to-one homomophism $BW : X(T) \to Pic(G/B)$. The image has finite index, equal to the order of the fundamental group of G.

In characteristic zero, a refinement of the above results leads to a decomposition of $K[G]$ as a sum of $G \times G$-modules. Indeed, the action $(G \times G) \times G \to G$, defined by $((g,h),x) \to gxh^{-1}$, defines a rational action of $G \times G$ on $K[G]$. The resulting decomposition of $K[G]$ into isotypic components leads to the following description of $K[G]$:

$$K[G] = \bigoplus_{\lambda \in X(T)_+} H^0(\lambda^*) \otimes H^0(\lambda).$$

The summands $H^0(\lambda^*) \otimes H^0(\lambda)$ are the **blocks** of $K[G]$ in the sense of Green [33]. Notice also that $K[G]$ has **simple $G \times G$-spectrum**. This is in fact one of the ways to define an affine **spherical variety**. See [76].

2.2.5 The Class Group of a Reductive Group

Let G be a connected, reductive group with coordinate algebra $K[G]$. In this section we calculate the class group of G in terms of certain extremal functions on G (using Proposition 2.19 and the Bruhat-Tits decomposition of G). Many of our results are contained explicitly or implicitly in [42], [75] and [137].

Let B and $B^- \subseteq G$ be opposite Borel subgroups of G. Let

$$T = B \cap B^-.$$

Then there is a **big cell**

$$BB^- \subseteq G.$$

BB^- is open and dense in G, and is isomorphic to $K^m \times (K^*)^n$ as varieties. In particular, $Cl(BB^-) = 0$. On the other hand,

$$G \backslash BB^- = \cup_{\alpha \in \Delta} \overline{Bs_\alpha B^-},$$

where Δ is the set of simple roots of T relative to B. Write

$$D_\alpha = \overline{Bs_\alpha B^-}.$$

We sometimes write $D_\alpha(G)$ if there is possibility of confusion. By part c) of Proposition 2.19, $Cl(G)$ is generated by $\{D_\alpha \mid \alpha \in \Delta\}$. If $f \in K[G]$ and

$$Z(f) \subseteq \cup_{\alpha \in \Delta} D_\alpha,$$

it follows easily that

$$BfB^- = K^* f.$$

Definition 2.50. *Let* $L(G) = \{ f \in K[G] \mid V(f) \subseteq \cup D_\alpha,\ f(1) = 1 \}$.

We refer to $L(G)$ as the **augmented cone** of G (although, strictly speaking, $L(G)$ is the set of lattice points of such a cone).

Define

$$c : L(G) \to Div(G)$$

by

$$c(f) = \sum_{\alpha \in \Delta} \nu_\alpha(f) D_\alpha$$

where ν_α is the valuation on $K[G]$ associated with the prime divisor $D_\alpha \subseteq G$ of G. Notice that c will not be injective unless G is a semisimple group.

Let

$$Div_0(G) = \oplus_{\alpha \in \Delta} \mathbb{Z} D_\alpha \subseteq Div(G).$$

Proposition 2.51. *$Cl(G) = Div_0(G)/ < c(L(G) >$, where $< c(L(G)) >$ is the subgroup of $Div_0(G)$ generated by $c(L(G))$. In particular, $Cl(G) = 0$ if and only if the ideal of each D_α is principal.*

Proof. By part c) of Proposition 2.19, $Cl(G)$ is generated by $\{D_\alpha\}$, while the principal divisors in $Div_0(G)$ are exactly the ones comming from $L(G)$.

Proposition 2.52. *There is a canonical one-to-one correspondence between $L(G)$ and the set $X(T)_+$ of dominant weights of irreducible representations of G.*

Proof. By Theorem 31.4 of [40], if $\lambda \in X(T)_+$ there is a function $c_\lambda \in L(G)$ such that the right G-submodule V_λ of

$$H^0(\lambda) = \{ f \in K[G] \mid f(xy) = \lambda(x) f(y) \text{ for all } x \in B^-, \ y \in G \}$$

generated by c_λ, is irreducible. It then follows from part b) of Theorem 2.48 that this V_λ is unique.

Conversely, any $c \in L(G)$ yields an irreducible representation V of G by considering the submodule of $K[G]$ generated by this c under right translation. By part a) of Theorem 2.48, $V = V_\lambda$ for some $\lambda \in X(T)_+$.

The coefficients $\{\nu_\alpha(f)\}$ in the formula $c(f) = \sum_{\alpha \in \Delta} \nu_\alpha(f) D_\alpha$ have the following interpretation for a semisimple group G. Let

$$\alpha^\vee = \frac{2\alpha}{(\alpha, \alpha)}$$

be the **coroot** associated with $\alpha \in \Delta$. Then, by Theorem 5.3 of [42],

$$\nu_\alpha(f) = (\alpha^\vee, \lambda) = < \alpha, \lambda >$$

where λ corresponds to f via Proposition 2.52. In particular, if $\lambda = \lambda_\alpha$ is fundamental and dominant, then $\nu_\alpha(\lambda_\beta)$ equals one if $\alpha = \beta$ and zero if $\alpha \neq \beta$.

Proposition 2.53. *Let G be connected and reductive, and let $G' = (G, G)$. Then*

a) $Cl(G) = Cl(G')$
b) In particular, the following are equivalent:
  i) $Cl(G) = 0$.
  ii) $Cl(G') = 0$.

Proof. Let Z be the connected center of G. The multiplication morphism $m : G' \times Z \to G$ is a central isogeny, as in §2 of [42]. Hence, by Proposition 2.6 of [42], there is an exact sequence

$$0 \to X(G) \to X(G' \times Z) \to X(ker(m)) \to Cl(G) \to Cl(G' \times Z) \to 0.$$

But $X(G) = X(G/G') = X(Z/(Z \cap G'))$ and $X(G' \times Z) = X(Z)$. Hence $X(G' \times Z) \to X(ker(m))$ is surjective, since $ker(m) = Z \cap G'$.

Proposition 2.54. *Let G be a connected reductive group. Then there exists a connected reductive group G_1, with $Cl(G_1) = 0$, and a finite dominant morphism $\pi : G_1 \to G$ with central kernel.*

Proof. By Proposition 1 of [75] (reproved in Corollary 3.3 of [42]), this is true for $G' = (G, G)$, which is semisimple. Say $f : \widehat{G'} \to G'$ is the universal cover of G'. Let Z be the connected center of G. Then the desired morphism is $g : \widehat{G'} \times Z \to G$, defined by $g(x, z) = f(x)z$.

Now let $L \subseteq G$ be a Levi factor of G. Then there exist opposite parabolic subgroups P, P^- of G such that $L = P \cap P^-$. However,

$$PP^- \cong U \times L \times U^-,$$

where $U = R_u(P)$ and $U^- = R_u(P^-)$. Since U and U^- are affine spaces, $Cl(L) = Cl(PP^-)$.

We conclude this section with the following corollary.

Corollary 2.55. *There exists a surjective morphism $Cl(G) \to Cl(L)$. In particular, if $Cl(G) = 0$ then $Cl(L) = 0$*

Proof. PP^- is open in G. Hence $Cl(G) \to Cl(PP^-)$ is surjective from part a) of Proposition 2.19.

2.2.6 Actions, Orbits, Invariants and Quotients

Let G be a reductive group, and let X be an irreducible variety. We assume that X is affine unless otherwise stated. An **action**

$$\mu : G \times X \to X$$

of G on X is a morphism of algebraic varieties such that:

a) for all $g, h \in G$ and $x \in X$, $\mu(g, \mu(h, x)) = \mu(gh, x)$,
b) for all $x \in X$, $\mu(1, x) = x$.

We denote $\mu(g, x)$ by gx. The **orbit** of $x \in X$ is $Gx = \{y \in X \mid y = gx \text{ for some } g \in G,\ x \in X\}$. The **isotropy subgroup** of $x \in X$ is $G_x = \{g \in G \mid gx = x\}$. An orbit $Gx \in X$ is **dense** if it is a dense subset of X in the Zariski topology. Any dense orbit is actually an open subset. The theory of algebraic monoids provides us with many important examples where some action $G \times X \to X$ has a dense orbit.

An orbit $Gx \in X$ is **closed** if it is a closed subset of X in the Zariski topology. Any orbit of minimal dimension is closed.

The action μ induces a linear action ρ of G on $K[X]$ as follows. For $g \in G$ and $f \in K[X]$ define $\rho_g(f) \in K[X]$ by $\rho_g(f)(x) = f(g^{-1}x)$ for all $x \in X$. ρ is **rational** in the sense that $K[X]$ is the union of its finite dimensional, G-stable subspaces. The **ring of invariants** $K[X]^G$ of μ (or ρ) is defined as follows:

$$K[X]^G = \{f \in K[X] \mid \rho_g(f) = f \text{ for all } g \in G\}.$$

The following result summarizes some of the fundamental theorems of **Geometric Invariant Theory**. The reader should consult [62, 65, 134] for an appreciation of the scope and significance of this theory.

Theorem 2.56. *Let $\mu : G \times X \to X$ be an action of the reductive group G on the affine variety X.*

a) $K[X]^G$ is a finitely generated K-algebra.
b) If we define the quotient *X/G to be the affine variety defined by $K[X]^G$, then the canonical morphism $\pi : X \to X/G$ identifies X/G with the set of closed orbits of G on X. In fact, the closure of any orbit Gx in X contains exactly one closed G-orbit.*

Notice that this notion of quotient is not usually an orbit space in the usual sense. But it has some categorical properties that are normally expected of any orbit space.

Example 2.57. Let $PGl_n(K) \times M_n(K) \to M_n(K)$ be the action defined by

$$(g, A) \to gAg^{-1}.$$

Then the quotient of this action can be identified as follows:
For $A \in M_n(K)$, let $det(tI - A) = t^n - \sigma_1(A)t^{n-1} + \cdots + \sigma_{n-1}(A)t + (-1)^n\sigma_n(A)$ be the characteristic polynomial of A. Then define

$$Ad : M_n(K) \to K^n$$

by $Ad(A) = (\sigma_1(A), \ldots, \sigma_n(A))$. This is our quotient in the sense of the above theorem. It is well known that the closure of the conjugacy class of A contains the semisimple part A_s of A. Furthermore, two semisimple endomorphisms are conjugate if and only if they have the same characteristic polynomial.

The following result is originally due to V. L. Popov [74].

Proposition 2.58. *Let X be a normal, irreducible, affine variety with trivial divisor class group. Assume the connected, semisimple, algebraic group G acts on X. Let X/G be the geometric invariant theory quotient of this action (as in Theorem 2.56). Then X/G is also a normal variety with trivial divisor class group.*

Proof. We let $A = K[X]$, so that $K[X/G] = K[X]^G$. We denote the action of G on elements $a \in A$ by $g(a)$. Let $a \in A^G$ be a non-unit. Then $a \in A$ is also a non-unit. Write $a = p_1p_2 \dots p_m$, where $p_i \in A$ is prime. Now for $g \in G$ we obtain $g(a) = g(p_1)\dots g(p_m)$. Since G is connected, the action stabilizes each irreducible component of $Z(\{a\})$. Thus for each i, $g(p_i) = \alpha_i(g)p_i$, for some unit $\alpha_i : G \to K^*$ with $\alpha_i(1) = 1$. But α_i must be constant since the unit group of $K[G]$ is K^*. Thus $\{p_i\} \subseteq A^G$. These p_i are easily seen to be prime in A^G. Thus A^G is a unique factorization domain.

2.2.7 Cellular Decompositions of Algebraic Varieties

Some of the well established ways to study the "topology" of an algebraic variety is the use of comparison theorems or base change theorems, along with results that tell us how to proceed when a variety can be broken up into manageable peices. Roughly speaking, a comparison theorem states that if an algebraic variety X is considered as a topological space X_{top} then the cohomology of X can be understood or calculated in terms of a more convenient cohomology theory. One of the most well known comparison theorems of this type states that, if $H^*(X, \mathbb{Q}_l)$ is the l-adic cohomology of the smooth, projective variety X, then

$$H^*(X, \mathbb{Q}_l) \otimes_{\mathbb{Q}_l} \mathbb{C} \cong H^*(X, \mathbb{C}).$$

A base change theorem usually concerns the situation when a variety X is subjected to some convenient base extension $X \to \overline{X}$. A lot of information about l-adic cohomology of X can be calculated in terms of the Weil zeta function of $\overline{X}$. This method of counting the points of the appropriate reduction mod p is particularly interesting in the theory of algebraic monoids. We are often interested in counting the elements of certain finite monoids $M(\mathbb{F}_{q^n})$ over the finite field $\mathbb{F}_{q^n}$. Letting $n \to \infty$ yields an interesting enumerative theory, as well as useful topological information about certain related algebraic varieties.

There is another method that applies to varieties that can be broken up into well-behaved peices, or cells. The most commonly studied cellular decompositions in algebraic geometry are those of Bialynicki-Birula [4]. If $S = K^*$ acts on a smooth complete variety X with finite fixed point set $F \subseteq X$, then $X = \bigsqcup_{\alpha \in F} X_\alpha$ where $X_\alpha = \{x \in X \mid \lim_{t \to 0} tx = \alpha\}$. Furthermore, X_α is isomorphic to an affine space. We refer to X_α as a *BB-cell.* If further, a reductive group G acts on X extending the action of S, we may assume (replacing S if necessary) that each X_α is stable under the action of some Borel subgroup B of G with $S \subseteq B$. In case X is a complete homogeneous space for G, each cell X_α turns out to consist of exactly one B-orbit.

But there are yet other types of cellular decompositions that do not arise from the method of [4] (as we shall see in Theorem 10.15), and these can also work out well homologically. In particular, let X be an irreducible algebraic variety, and assume that X is a disjoint union

$$X = \bigsqcup_i C_i$$

of *cells*, where each cell C_i is isomorphic to the affine space K^{n_i}. Assume further that $\cup_{n_i \leq m} C_i$ is closed in X for each $m > 0$.

Theorem 2.59. *The natural map*

$$c_X : A_*(X) \to H_*(X, \mathbb{Z})$$

from the Chow ring of X to cellular homology, is an isomorphism. Furthermore, $\{\overline{C_i} \mid n_i = m\}$ is a $\mathbb{Z}$-basis for $A_m(X)$.

Proof. See Fulton [30] Example 1.9.1 and Example 19.1.11.

2.3 Semigroups

The purpose of this section is to assemble some of the basic ideas from semigroup theory that are particularly relevant to the theory of algebraic monoids. In each situation, we try to illustrate the material with relevant examples from linear algebra.

2.3.1 Basic Semigroup Theory

A set S together with an associative operation $m : S \times S \to S$ is called a **semigroup**. If S has an element $1 \in S$ such that $1s = s1 = s$ for all $s \in S$, then S is called a **monoid**. If S is a semigroup, we define $S^1 = S$ if S is a monoid, and $S^1 = S \cup \{1\}$ with the obvious multiplication, if S is not a monoid. In either case S^1 is a monoid. If $X \subseteq S$ then $E(X) = \{e \in X | e^2 = e\}$ is the set of **idempotents** of X. If S, T are semigroups, then a map $\psi : S \to T$ is a **homomorphism** if $\psi(xy) = \psi(x)\psi(y)$ for all $x, y \in S$. The equivalence relation on S induced by a homomorphism is called a **congruence**. A subsemigroup of S which is a group is called a **subgroup** of S. Notice that the identity element of a subgroup of S could be any idempotent of S. If $e \in S$ is an idempotent, then the unit group of eSe is a maximal subgroup of S. All maximal subgroups of S are obtained this way. An **ideal** of S is a nonempty subset J of S such that if $x \in J$ then $S^1 x S^1 \subseteq J$. There is also the notion of one-sided ideal. If S has a minimum ideal K, it is called the **kernel** of S. Any finite semigroup has a kernel.

An element $a \in S$ is **regular** if $axa = a$ for some $x \in S$. S is **regular** if each of its elements is regular. Let M be a monoid with unit group G. We say that M is **unit regular** if, for each $a \in M$, there is a unit $g \in G$ such that $a = aga$. Equivalently, $M = GE(M) = E(M)G$. The monoid $M_n(K)$ of $n \times n$ matrices is unit regular, and the semigroup S of singular $n \times n$ matrices is regular.

Let S be a semigroup, and let $M = S^1$. It is useful to introduce *Green's relations* [32].

Definition 2.60. *Let $a, b \in M$.*

a) $a\mathcal{R}b$ if $aM = bM$.
b) $a\mathcal{L}b$ if $Ma = Mb$.
c) $a\mathcal{J}b$ if $MaM = MbM$.
d) $a\mathcal{H}b$ if $a\mathcal{R}b$ and $a\mathcal{L}b$.
e) $a\mathcal{D}b$ if $a\mathcal{R}c$ and $c\mathcal{L}b$ for some $c \in M$.

We denote by H_x (or H if no confusion is possible) the $\mathcal{H}$-class of x: and similarily for $\mathcal{R}$, $\mathcal{L}$ and $\mathcal{J}$.

Example 2.61. Let $M = M_n(K)$. $a\mathcal{L}b$ if and only a and b are row equivalent. $a\mathcal{R}b$ if and only if a and b are column equivalent. $a\mathcal{J}b$ if and only if rank(a)=rank(b). In this example $\mathcal{J} = \mathcal{D}$.

Remark 2.62. Let S be a semigroup.

a) If $a \in S$ then a lies in a subgroup of S if and only if $a\mathcal{H}e$ for some idempotent $e \in S$.
b) If $a \in S$, $e \in E(S)$, $a\mathcal{R}e$ and H is the $\mathcal{H}$-class of e, then Ha is the $\mathcal{H}$-class of a.
c) For $e, f \in E(S)$, $e\mathcal{R}f$ if and only if $ef = f$ and $fe = e$.
d) Let $a \in S$ be a regular element. Then $a = axa$ for some $x \in S$. Then $e = ax$, $f = xa \in E(S)$, and $e\mathcal{R}a\mathcal{L}f$. Thus a is regular if and only if $e\mathcal{R}a$ for some $e \in E(S)$ if and only if $a\mathcal{L}f$ for some $f \in E(S)$.

2.3.2 Strongly π-regular Semigroups

We begin this section with a definition.

Definition 2.63. *Let S be a semigroup. We say that S is* strongly π-regular *($s\pi r$) if for any $x \in S$, $x^n \in H_e$ for some $e \in E(S)$ and some $n > 0$.*

Remark 2.64. a) $s\pi r$ is the main notion that best captures the semigroup theoretic essence of many linear semigroups. Indeed, if $S = M_n(K)$, then any $x \in S$ can be written uniquely as $x = r + n$, where n is nilpotent, rank(x^m)=rank(x) for any $m > 0$, and $rn = nr = 0$ (*Fitting decomposition*). Then $x^m \in H_e$ where e is the unique idempotent of S with the same rank as r, such that $er = re = r$.

b) More generally, let S be an $s\pi r$ subsemigroup of the semigroup T (for example $T = M_n(K)$) with $a \in S$ and $e \in E(T)$. If $a\mathcal{H}e$ in T, then $e \in S$ and $a\mathcal{H}e$ in S.
c) Any finite semigroup is $s\pi r$.

The following elementary result is taken from [82]. We include the proof for convenience. This should indicate the usefulness of the $s\pi r$ condition.

Theorem 2.65. *Let S be an $s\pi r$ semigroup, $a, b, c \in S$. Then*

a) $a\mathcal{J}ab$ implies $a\mathcal{R}ab$, and $a\mathcal{J}ba$ implies $a\mathcal{L}ba$.
b) $ab\mathcal{J}b\mathcal{J}bc$ implies $b\mathcal{J}abc$.
c) If $e \in E(S)$, J is the $\mathcal{J}$-class of e and H is the $\mathcal{H}$-class of e, then $J \cap eSe = H$.
d) $\mathcal{J} = \mathcal{D}$ on S.
e) If $a\mathcal{J}a^2$ then the $\mathcal{H}$-class of a is a group.
f) $a\mathcal{J}ab\mathcal{J}b$ if and only if $a\mathcal{L}e\mathcal{R}b$ for some $e \in E(S)$.
g) Any regular subsemigroup of S is an $s\pi r$ semigroup.

Proof. For a) suppose that $a\mathcal{J}ab$. Then $xaby = a$ for some $x, y \in S^1$. Then $x^i a(by)^j = a$, for all $i, j > 0$. But there exists $j > 0$ such that $(by)^j\mathcal{H}e$ for some $e \in E(S)$. Then $a = ae \in a(by)^jS \subseteq abS$. Hence $a\mathcal{R}ab$. For b) we first get $ab\mathcal{L}b$ from a). Then $abc\mathcal{L}bc\mathcal{J}b$. For c) let $a \in eSe \cap J$. Then by a), $e\mathcal{R}ea = a = ea\mathcal{L}e$. Then $e\mathcal{H}a$. For d), let $a, b \in S$ be such that $a\mathcal{J}b$. Then there exist $x, y \in S$ such that $xay = b$. So $a\mathcal{J}xa\mathcal{J}xay = b$. Then again by a), $a\mathcal{L}xa\mathcal{R}b$. Thus $x\mathcal{D}y$. For e), let H denote the $\mathcal{H}$-class of a. From a), we see that $a\mathcal{H}a^2$. Then $a^2x = a$ for some $x \in S^1$. Then $a^{i+1}x^i = a$ for all $i > 0$. Thus $a^i\mathcal{R}a$ for all $i > 0$. By a) again, $a^i \in H$ for all $i > 0$. But there exist $j > 0$ and $e \in E(S)$ such that $a^j\mathcal{H}e$. But then $e \in H$ and so H is a group. For f), suppose that $a\mathcal{J}ab\mathcal{J}b$. Then by a), $a\mathcal{R}ab\mathcal{L}b$. Hence there exist $x, y \in S^1$ such that $abx = a$ and $yab = b$. Then $ya = yabx = bx$. Hence $aya = a$ and $bxb = b$. Thus $ya \in E(S)$ and $a\mathcal{L}ya = bx\mathcal{R}b$. Conversely, assume that there exists $e \in E(S)$ such that $a\mathcal{L}e\mathcal{R}b$. Thus $xa = by = e$ for some $x, y \in S$. Hence $ab|xaby = e|a|ab$. Thus $a\mathcal{J}ab$. For g), Let $a \in S'$. There exists $i > 0$ and $e \in E(S)$ such that $b = a^i\mathcal{H}e$ in S. But there exists $x \in S'$ such that $b^2xb^2 = b^2$. Then $bxb = e$, and so $e \in E(S')$ and $b\mathcal{H}e$ in S'.

Definition 2.66. *Let S be an $s\pi r$ semigroup. A $\mathcal{J}$-class J of S is* regular *if $E(J) \neq 0$. Equivalently, every element of J is regular. Let $\mathcal{U}(S)$ denote the partially ordered set of all regular $\mathcal{J}$-classes of S. Let $J \in \mathcal{U}(S)$, and define $J^0 = J \cup \{0\}$, with multiplication*

$$xy = \begin{cases} 0 & , \text{ if } x = 0,\ y = 0 \text{ or } xy \notin J \\ xy & , \text{ if } xy \in J\,. \end{cases}$$

Definition 2.67. *a) A* completely simple semigroup *is an $s\pi r$ semigroup with no ideals other than S.*

b) A completely 0-simple semigroup *is an $s\pi r$ semigroup with no ideals other than $\{0\}$ and S.*

Remark 2.68. a) Definition 2.67 is not the standard definition of simple and completely simple semigroups. However, by a theorem of Munn [63], our definitions are equivalent to the standard ones.

b) Let S be an $s\pi r$ semigroup, $J \in \mathcal{U}(S)$. If $a, b \in J$, then there exist $s, t, x \in S^1$ such that $sat = b$ and $axa = a$. Then $b = (sax)a(xat) \in JaJ$. Thus J^0 is completely 0-simple semigroup.

c) Let S be $s\pi r$, $J \in \mathcal{U}(S)$. If $E(J)^2 \subseteq J$, then by Theorem 2.65b) $J^2 = J$, and hence J is completely simple.

d) A completely 0-simple semigroup has two $\mathcal{J}$-classes, while a completely simple semigroup has one $\mathcal{J}$-class.

It turns out that there is a very satisfying structure theorem for completely simple and completely 0-simple semigroups.

Definition 2.69. *Let G be a group and let Γ, Λ be non-empty sets.*

a) Let $P : \Lambda \times \Gamma \to G$ be any map of sets. Define $S = \Gamma \times G \times \Lambda$, with multiplication

$$(i, g, j)(k, h, l) = (i, gP(j, k)h, l).$$

One checks that S is a completely simple semigroup.

b) Let $P : \Lambda \times \Gamma \to G \cup \{0\}$ be any map of sets such that for all $i \in \Gamma$ there exists $j \in \Lambda$ with $P(j, i) \neq 0$. Define $S = (\Gamma \times G \times \Lambda) \cup \{0\}$ with multiplication

$$(i, g, j)(k, h, l) = \begin{cases} 0 & , \text{ if } P(j, k) = 0 \\ (i, gP(j, k)h, l) & , \text{ if } P(j, k) \neq 0 \,. \end{cases}$$

One checks that S is a completely 0-simple semigroup.

In case a), S is called a Rees matrix semigroup without zero over *G with* sandwich matrix *P. In case b) S is called a* Rees matrix semigroup with zero over *G with* sandwich matrix *P.*

The following theorem is due to D. Rees [15].

Theorem 2.70. *a) Any completely simple semigroup is isomorphic to a Rees matrix semigroup without zero.*

b) Any completely 0-simple semigroup is isomorphic to a Rees matrix semigroup with zero.

Proof. We sketch the proof of b). Part a) follows from this, since we can construct a completely 0-simple semigroup from a completely simple semigroup by adjoining a superfluous zero element.

So let S be completely 0-simple semigroup. Thus $S = J \cup \{0\}$, where $J \subseteq S$ is a regular $\mathcal{J}$-class of S. Choose $e \in E(S)$, and let H, L, R be the $\mathcal{H}$-class, $\mathcal{L}$-class, $\mathcal{R}$-class of e, respectively. Define

$$\Gamma = L/\mathcal{R} = L/\mathcal{H},\ \Lambda = R/\mathcal{L} = R/\mathcal{H}.$$

For $\lambda \in \Lambda$ choose $r_\lambda \in \lambda$, and for $\gamma \in \Gamma$ choose $l_\gamma \in \gamma$. Define $P : \Lambda \times \Gamma \to S$ by

$$P(\lambda, \gamma) = r_\lambda l_\gamma.$$

By part a) of Theorem 2.65 we see that, if $r_\lambda l_\gamma \neq 0$, then $r_\lambda l_\gamma \in H$. Thus we have a map $P : \Lambda \times \Gamma \to H \cup \{0\}$. One can check, as in Theorem 1.9 of [82], that P is a sandwich matrix, and that S is isomorphic to the Rees matrix semigroup $S' = (\Gamma \times H \times \Lambda) \cup \{0\}$ with sandwich matrix P. In fact, an isomorphism

$$\psi : S' \to S$$

is given by $\psi(0) = 0$ and $\psi(\gamma, h, \lambda) = r_\lambda h l_\gamma$. One must check that ψ is well defined and bijective.

2.3.3 Special Types of Semigroups

In this section we introduce some of the different types of semigroups that show up in the theory of algebraic semigroups. Certainly we are not intending to be encyclopedic on this point.

Definition 2.71. *A* semilattice *is a commutative semigroup consisting of idempotents.*

Definition 2.72. *a) Let S be a semigroup, and assume that $S = \sqcup_{\alpha\in\Omega} S_\alpha$ is partitioned into a disjoint union of subsemigroups. Then we say that S is a* semilattice *of the S_α if, for all $\alpha, \beta \in \Omega$, there exists $\delta \in \Omega$ such that $S_\alpha S_\beta \cup S_\beta S_\alpha \subseteq S_\delta$.*

b) A semigroup S is completely regular *if it is the union of its subgroups.*

c) A semigroup S is a semilattice of groups *if, in addition to being completely regular, each of its $\mathcal{J}$-classes is in fact an $\mathcal{H}$-class.*

It turns out that the semigroup S is completely regular if and only if it is a semilattice of completely simple semigroups.

Example 2.73. a) Let S be the set of diagonal $n \times n$ matrices. Then S is a semilattice of groups.

b) Let S be the set of upper-triangular $n \times n$ matrices $A = (a_{i,j})$, of rank $n - 1$, such that $a_{n,n} = 0$. Then S is completely simple.

Let S be a semigroup and $a, b \in S$. Recall that a **divides** b, written $a|b$, if $xay = b$ for some $x, y \in S^1$.

Definition 2.74. *A semigroup S is* archimedean *if, for all $a, b \in S$, $a|b^i$ for some $i > 0$.*

Remark 2.75. a) A semigroup S is a semilattice of archimedean semigroups if and only if, for all $a, b \in S$, $a|b$ implies $a^2|b^i$ for some $i > 0$. For such a semigroup, define $x \sim y$ in S if $x|y^i$ for some $i > 0$ and $y|x^j$ for some $j > 0$. Then $\sim$ is the desired semilattice decomposition. See Theorem 1.15 of [82] for more details.

b) $T_n(K)$, the monoid of upper-triangular matrices, is a semilattice of archimedean semigroups. The corresponding semilattice in this case is canonically isomorphic to the semilattice of diagonal idempotents of $T_n(K)$.

c) Any commutative semigroup is a semilattice of archimedean semigroups.

Definition 2.76. *A semigroup S is called an* inverse semigroup *if for each $x \in S$ there exists a unique $x^* \in S$ such that*

$$xx^*x = x, \text{ and } x^*xx^* = x^*$$

Example 2.77. Let N be the set of $n \times n$ matrices with at most one nonzero entry in each row or column. Then N is an inverse semigroup.

Certain finite inverse semigroups will play the rôle of the Weyl group in the theory of algebraic monoids. See Proposition 8.1 and Theorem 8.8.

2.4 Exercises

2.4.1 Abstract Semigroups

1. Check that the definitions of S in 2.69 in fact yield completely (0-)simple semigroups.
2. Let $S = \Gamma \times G \times \Lambda$ be a completeley simple semigroup with sandwich matrix $P : \Lambda \times \Gamma \to G$. Identify the Green's relations $\mathcal{R}, \mathcal{L}$ and $\mathcal{H}$ on S in terms of P.
3. Prove that $T_n(K)$ is a semilattice of archimedean semigroups.
4. Prove that $M_n(K)$ is $s\pi r$.
5. Prove that $S = \{x \in M_n(K) \mid rank(x) \leq 1\}$ is a completely 0-simple semigroup.

3

Algebraic Monoids

The theory of algebraic monoids is built on the theory of algebraic groups, the theory of torus embeddings, and related semigroup constructions. Indeed, if M is an irreducible, algebraic monoid, then $M = \overline{G}$ where G is the algebraic group of units of M. So we hope the reader can acquire some familiarity with algebraic groups, Lie algebras, Tits buildings and torus embeddings [7, 40, 140, 31]. A brief summary of some of this essential background was assembled in the previous chapter.

In this chapter we acquaint the reader with some of the basic results about irreducible monoids. A significant number of these results amount to marrying the semigroup concepts with the geometric concepts. Where appropriate, we sketch some proofs. Nearly everything in this section is discussed in more detail in Putcha's monograph [82]. We reproduce some of Putcha's results for the convenience of the reader.

The reader might wish to begin with a more concrete and combinatorial approach, with many useful examples. In that case, he should consult Solomon's survey [128]. On the other hand, he might wish to start from the point of view of linear semigroups, in which case he should consult Okninski's book [66].

3.1 Linear Algebraic Monoids

Definition 3.1. *Let K be an algebraically closed field.*

a) An linear algebraic monoid *M is an affine, algebraic variety together with an associative morphism $\mu : M \times M \to M$ and an identity element $1 \in M$ for μ.*
b) M is irreducible *if it cannot be expressed as the union of two, proper, closed, non-empty subsets.*
c) The irreducible components *of M are the maximal, irreducible subsets of M.*

We often write “algebraic monoid” when we mean “linear algebraic monoid”. This abuse of language should not cause problems, since we are here concerned only with linear monoids, and it is easily seen that these are all affine. The problem of characterizing affine algebraic monoids among algebraic monoids has been discussed in [101, 121].

If M is an algebraic monoid, there is a unique, irreducible component $M^0 \subseteq M$ such that $1 \in M^0$. Indeed, $M^0 = \overline{G^0}$ where G^0 is the identity component of the unit group G of M. G is an algebraic group, open in M. Notice also that the monoid structure on an irreducible algebraic monoid is uniquely determined by the group structure of its unit group. This is so because the group is open and dense in the monoid. See Proposition 3.12 below.

An algebraic monoid M may be identified by its *bialgebra* $A = K[M]$.

Definition 3.2. *A* bialgebra *is a K-algebra A together with a coassociative morphism $\bigtriangledown : A \longrightarrow A \otimes_K A$ and counit $\varepsilon : A \to K$.*

One obtains an algebraic monoid $M = M(A)$ from the bialgebra A as follows. Define

$$M = Hom_{K-alg}(A, K).$$

The multiplication $*$ on M is defined via $\bigtriangledown$:

$$f * g = (f \otimes_K g) \circ \bigtriangledown.$$

For example consider the bialgebra $A = K[T_{ij} \mid 1 \le i, j \le n]$ with

$$\bigtriangledown : A \longrightarrow A \otimes_K A \quad \text{defined by}$$
$$\bigtriangledown (T_{ij}) = \sum_h T_{ih} \otimes T_{hj}\ , \qquad \text{and}$$
$$\varepsilon : A \longrightarrow K \quad \text{defined by}$$
$$\varepsilon(T_{ij}) = \begin{cases} 1\ , & i = j \\ 0\ , & i \ne j \end{cases} .$$

One easily checks that A is the coordinate ring of $M_n(K) = Hom_{K-alg}(A, K)$.

The bialgebra approach is useful technically in decomposing the coordinate algebra into blocks. See § 9.4.

Definition 3.3. *A* morphism φ *of algebraic monoids M and N is a morphism $\varphi : M \to N$ of algebraic varieties such that $\varphi(xy) = \varphi(x)\varphi(y)$ for $x, y \in M$, and $\varphi(1) = 1$.*

For an interesting example, consider $\varphi : M_2(K) \longrightarrow M_3(K)$ defined by

$$\varphi \begin{pmatrix} a & b \\ c & d \end{pmatrix} = \begin{pmatrix} a^2 & ab & b^2 \\ 2ac & ad + bc & 2bd \\ c^2 & cd & d^2 \end{pmatrix} .$$

Definition 3.4. *Let M be an algebraic monoid.*

a) $E(M) = \{e \in M \mid e^2 = e\}$ *is the set of* idempotents *of* M.
b) M *is* regular *if for each* $a \in M$ *there exists* $x \in M$ *such that* $axa = a$.
c) M *is* unit regular *if* $M = GE(M)$.
d) $\mathfrak{U}(M)$ *is the set of subsets of* M *of the form* GeG*, where* $e \in E(M)$*. These subsets are known as* regular $\mathfrak{J}$*-classes. This is not the usual definition, but is equivalent for irreducible algebraic monoids.*

It turns out that any regular, irreducible monoid is unit regular. See [82].

Example 3.5. Let $M = M_n(K)$. This is one of the motivating examples for much of the important structure theory of reductive monoids. Notice that $M_n(K)$ is regular.

Example 3.6. **(Finite monoids)** Let M be a finite monoid, and let $A = Hom(M, K)$. Define $\bigtriangledown : A \longrightarrow A \otimes A = Hom(M \times M, K)$ by the rule $\bigtriangledown(f)(s,t) = f(st)$. It is easy to check that $(A, \bigtriangledown)$ is a bialgebra and that the associated algebraic monoid is canonically isomorphic to M.

Example 3.7. **(Semidirect products)** Let M and N be algebraic monoids and suppose that we have a morphism $\gamma : M \times N \longrightarrow N$ of algebraic varieties such that

$$\gamma(s, t_1t_2) = \gamma(s, t_1)\gamma(s, t_2) \text{ for } s \in M, t_1, t_2 \in N$$

and

$$\gamma(s_1s_2, t) = \gamma(s_1, t)\gamma(s_2, t) \text{ for } s_1, s_2 \in M, t \in N \ .$$

Write t^s for $\gamma(s,t)$. Then $M \times N$ is an algebraic monoid with $(s_1, t_1)(s_2, t_2) = (s_1s_2, t_1^{s_2}t_2)$.

Any algebraic monoid M is strongly π-regular, as in § 2.3.2. Precisely, if $x \in M$ then $x^n \in H_e$, the unit group of eMe, for some $e \in E(M)$. Furthermore, $e \in \{x^n \mid n > 0\}$. The integer n can be chosen independently of $x \in M$. In any case, there is always an abundance of idempotents. For example, if $E(M) = \{1\}$, then $M = G$. One can easily check this condition for $M = M_n(K)$. In general, one can use the following theorem.

Theorem 3.8. *Let M be an algebraic monoid. Then for some $n > 0$, there exists a morphism $\rho : M \longrightarrow M_n(K)$, of algebraic monoids, such that ρ is a closed embedding of algebraic varieties.*

The basic idea behind Theorem 3.8 is **right translation of functions**. Define

$$\gamma : M \longrightarrow End(K[M])$$

by the rule $\gamma_s(f)(t) = f(ts)$, where $s, t \in M$, and $f \in K[M]$. Then γ is a morphism of monoids. To obtain the morphism ρ, one chooses a finite-dimensional subspace $V \subseteq K[M]$ such that

a) $\gamma_s(V) \subseteq V$ for all $s \in M$.
b) V generates $K[M]$ as a K-algebra.

It then turns out that $\rho : M \longrightarrow End(V)$, $\rho(s)(v) = \gamma_s(v)$, satisfies the conclusion of Theorem 3.8.

Corollary 3.9. *Let $\varphi : M \to N$ be a morphism of algebraic monoids. If $e \in E(N) \cap \varphi(M)$, then there exists $f \in E(M)$ such that $\varphi(f) = e$.*

Proof. By Theorem 3.8 we can think of M as a submonoid of some $End(V)$. But then $\varphi^{-1}(e)$ is a closed subsemigroup of $End(V)$. By the comment just preceding Theorem 3.8, any such semigroup has idempotents.

Let M be an algebraic monoid. Recall that an **ideal** $I \subseteq M$ is a nonempty subset such that $MIM \subseteq I$.

Theorem 3.10. *Let M be an irreducible monoid. If $P \subseteq M$ is a* prime *ideal (so that $M \backslash P$ is multiplicatively closed) then P is closed in M. Furthermore, there exists a morphism $\chi : M \to K$ of algebraic monoids such that $P = \chi^{-1}(0)$.*

Theorem 3.10 is proved in [101]. The strategy there is to first show that the result is true for irreducible monoids Z with unit group a torus. The general case then follows once it is shown that P is determined by $P \cap \overline{T}$. See Exercises 8, 9, 10 and 11 of § 3.5.3 for an outline of this proof.

It follows from Theorem 3.10 that any quasi-affine, irreducible monoid is actually affine.

One should notice that the unit groups of irreducible algebraic monoids are "big". The next result records how this observation is reflected in Green's relations (Definition 2.60).

Proposition 3.11. *Let M be an irreducible algebraic monoid. Let $a, b \in M$, $e, f \in E(M)$ and $G = G(M)$.*

a) $a\mathfrak{R}b$ if and only if $aG = bG$.
b) $a\mathfrak{L}b$ if and only if $Ga = Gb$.
c) $a\mathfrak{J}b$ if and only if $GaG = GbG$.

The main point behind the proofs of a), b) and c) is the following basic fact from algebraic group theory: if $G \times X \to X$ is a regular action then any orbit is open in its closure.

If M is an algebraic monoid, then $\mathfrak{J} = \mathfrak{D}$. Indeed, by Theorem 1.4 of [82], this is true for any $s\pi r$ monoid.

Proposition 3.12. *Let M be an algebraic monoid and let $e \in E(M)$. Then H_e, the unit group of eMe, is an algebraic group, open in eMe.*

Proof. This follows from Theorem 3.8 applied to eMe. Indeed, $\rho(M) \cap G\ell_n(K) = \rho(G(M))$, which is closed in $G\ell_n(K)$ and open in $\rho(M)$.

We end this section with a statement of the fundamental results about idempotents. See Corollary 6.4 and Corollary 6.8 of [82] for proofs.

Proposition 3.13. *Let M be an irreducible algebraic monoid, and let $e \in E(M)$. Then*

a) $e \in \overline{T}$ *for some maximal torus* $T \subseteq G$. *So* $E(M) = \bigcup_{g \in G} gE(\overline{T})g^{-1}$.

b) $e\mathfrak{J}f$ *if and only if* $g^{-1}eg = f$ *for some* $g \in G$.

c) $e\mathfrak{R}f$ *if and only if* $g^{-1}eg = f = eg$ *for some* $g \in G$.

d) $e\mathfrak{L}f$ *if and only if* $geg^{-1} = f = ge$ *for some* $g \in G$.

It is possible to characterize which irreducible algebraic groups G can occur non trivially as the unit group of some algebraic monoid.

Theorem 3.14. *Let G be an irreducible algebraic group. Then the following are equivalent.*

a) There exists an irreducible, algebraic monoid M with unit group G such that $G \neq M$.

b) $X(G) \neq \{1\}$.

See Execise 6 of 3.5.1 for an outline of the proof.

3.2 Normal Monoids

Unlike the case of algebraic groups, not every irreducible algebraic monoid is normal as an algebraic variety. This is mainly a technical nuisance, since any algebraic monoid has a lot in commom with its normalization. On the other hand, there are important advantages to normal monoids. This will become apparent in the classification problem. See Theorem 5.2.

Let X be an irreducible, algebraic variety. Recall that the **normalization** $\eta : X' \to X$ of X is the unique finite, birational morphism η from an irreducible normal variety X' to X. If X is affine, the coordinate algebra of $K[X']$ is the integral closure of $K[X]$ in its field of fractions.

Luckily we have the following result for the normalization of an algebraic monoid. This was first recorded in [100].

Proposition 3.15. *Let M be an irreducible, algebraic monoid with unit group G. Let $\eta : M' \to M$ be the normalization of M. Then M' has the unique structure of an algebraic monoid such that η is a finite, birational morphism of algebraic monoids.*

Proof. One must check that the multiplication morphism $m : M \times M \to M$ extends to a morphism $m' : M' \times M' \to M'$. This extension is possible because of the universal property of normalization.

Remark 3.16. If M is an irreducible, algebraic monoid with unit group G, and $\pi : G' \to G$ is a finite dominant morphism, define M' so that $K[M']$ is the integral closure of $K[M]$ in $K[G']$. Then M' is an irreducible algebraic monoid with unit group G'. Furthermore, there is a unique (finite) morphism $\theta : M' \to M$ such that $\theta|G' = \pi$.

Example 3.17. Let $M = \{(x, y) \mid x^2 = y^3\}$. Then M is an irreducible, algebraic monoid with unit group isomorphic to K^*, and pointwise multiplication. Here

$$K[M] = K[X,Y]/(X^2 - Y^3),$$

which is not normal. The normalization of M is

$$\eta : K \to M,$$

defined by $\eta(t) = (t^3, t^2)$.

3.3 D-monoids

The closure of a maximal torus plays a special rôle in the theory of algebraic monoids. For example, from Proposition 3.13, we see that any idempotent is in the closure of a maximal torus. This is good news because such monoids have been much studied as **torus embeddings**. In this section we give a short description of this class of monoids.

Example 3.18. Let H be a closed, connected subgroup of $D_n(K)^* \cong K^* \times \cdots \times K^*$. Let $M = \overline{H}$, the Zariski closure of H in $D_n(K)$, the set of diagonal matrices. Then M is a semilattice of groups in the sense of Definition 2.72 c). Such monoids are called **D-monoids**. Furthermore, $\mathcal{U}(M) = \{[x] \in \overline{H} \mid [x] = [y]$ if there exists $g \in H$ such that $y = gx\}$. This is just another way of talking about affine torus embeddings. $\mathcal{U}(M)$ is isomorphic to the face lattice of a rational polytope [31].

D-monoids are described axiomatically as follows.

Definition 3.19. *A* D-monoid *is an irreducible, algebraic monoid* M *such that* $K[M]$ *is spanned over* K *by* $X(M) = \{\chi \in K[M] \mid \bigtriangledown(\chi) = \chi \otimes \chi\}$. $X(M)$ *is called the* character monoid *of* M. *Notice that we may regard* $X(M)$ *as a subset of* $X(G)$.

It is easy to see that M is a D-monoid if and only if M is isomorphic to a closed submonoid of $D_n(K)$ for some $n > 0$. Indeed, choose characters $\chi_1, \ldots, \chi_n \in X(M)$ that generate $K[M]$ as an algebra, and define

$$\psi : M \to D_n(K)$$

by $\psi(z) = (\chi_1(z), \ldots, \chi_n(z))$.

If G is a D-group and $\rho : G \to D_n(K) \subseteq M_n(K)$ is a rational representation of G, the set of **weights** of ρ is the set $\Phi = \{\chi \in X(G) \mid \rho(g)(v) = \chi(g)v$ for all $g \in G$ and some nonzero $v \in K^n\}$.

Proposition 3.20. *Let M be a D-monoid with unit group G. Let*

$$\rho : G \to D_n(K)$$

be a representation of G with weights $\Phi \subseteq X(G)$. Then ρ extends to a representation $\overline{\rho} : M \to D_n(K)$ if and only if $\Phi \subseteq X(M)$.

Proof. It is easy to check that $\rho^*(K[D_n(K)]) \subseteq K[G]$ is the subalgebra generated by Φ.

Remark 3.21. As we have already mentioned, if Z is a D-monoid with unit group T, then the set of T-orbits on Z is naturally the face lattice of a rational polytope $\mathcal{P}(Z)$. But each T-orbit H of Z contains exactly one idempotent $e = e(H)$. So the face lattice of $\mathcal{P}(Z)$ is in one-to-one correspondence with $E(Z)$, the set of idempotents of Z. The order relation "$e \leq f$, if $ef = fe = e$" on $E(Z)$ corresponds to the lattice ordering on the face lattice of $\mathcal{P}(Z)$.

One can also identify the face lattice of $\mathcal{P}(Z)$ as the semilattice of archimedean components of the commutative monoid $X(Z)$. See Remark 2.75 c).

Many useful properties of rational polytopes transfer over to the semilattice $E(Z)$. Recall that a **ranked poset** is a poset P with a rank function $r : P \to \mathbb{N}$ such that, if x covers y, then $r(x) = r(y) + 1$. This is another way of saying that all the maximal chains of P have the same length. Some authors refer to a ranked poset as a graded poset.

Proposition 3.22. *Let Z be a D-monoid of dimension n.*

a) $E(Z)$ is a ranked poset. The rank function here is $r(e) = dim(Te)$.
b) If $f \in E(Z)$, let $E^1(f) = \{e \in E(Z) \mid r(e) = n - 1\}$. Then

$$f = \Pi_{e \in E^1(f)} e.$$

Proof. The proof amounts to a translation of well known properties of polytopes to the language of D-monoids.

It is of interest to know when a D-monoid Z is normal. Also it is of interest to identify the coordinate ring of the normalization (Proposition 3.15) of Z.

Proposition 3.23. *Let Z be an irreducible D-monoid with unit group G and coordinate ring the monoid algebra $K[X(Z)]$. Observe that $X(Z) \subseteq X(G)$. The following are equivalent.*

a) Z is normal.
b) If $\chi \in X(G)$ and $\chi^m \in X(Z)$ for some $m > 0$, then $\chi \in X(Z)$.

Let $S = \{\chi \in X(G) \mid \chi^m \in X(Z)$ for some $m > 0\}$. Then, if $\eta : Z' \to Z$ is the normalization of Z, the coordinate ring of Z' is the semigroup algebra of S.

We leave the proof to the reader.

See Example 3.17 for a simple example. Notice that the normalization morphism induces a bijection on idempotents.

3.4 Solvable Monoids

Definition 3.24. *An irreducible monoid M is* solvable *if $G(M)$ is a solvable, algebraic group.*

Remark 3.25. By Theorem 17.6 of [40] and Theorem 3.8 above, any solvable monoid M is isomorphic to a closed submonoid of $T_n(K)$, the upper triangular monoid, for some $n > 0$. Using this fact one can construct a universal morphism $\pi : M \to Z$, to a D-monoid, such that $\pi|\overline{T} : \overline{T} \to Z$ is an isomorphism for any maximal torus $T \subseteq G = G(S)$. If M is irreducible and $B \subseteq G$ is a Borel subgroup, then $\overline{B} \subseteq M$ is solvable. Furthermore $B\overline{B} \subseteq \overline{B}$. Thus by a Theorem of Steinberg [134] $M = G\overline{B}$. Similarly, $M = \bigcup_{g \in G} g\overline{B}g^{-1}$.

The following result is due to Putcha; Corollary 6.32 of [82].

Theorem 3.26. *Let M be an irreducible monoid with zero and unit group G. Then the following are equivalent:*

a) G is solvable.
b) M is a semilattice of archimedean semigroups.

We refer the reader to [82] for the proof. Notice however that, from Remark 3.25 above, it suffices to prove that $T_n(K)$ is a semilattice of archimedean semigroups. See Excercise 3 of 2.4.1. Putcha obtains other interesting characterizations of solvable monoids. See Theorem 6.35 of [82] for example.

3.5 Excercises

3.5.1 Linear Algebraic Groups

1. Let G be an algebraic group acting on a variety X. Suppose that $Y \subseteq X$ is closed and $gY \subseteq Y$ for some $g \in G$. Prove that $gY = Y$. Hint: $Y \supseteq gY \supseteq g^2Y \ldots$, yet each g^sY is closed in Y.
2. Let $U \subseteq G$ be a closed subsemigroup, where G is an algebraic group.
   a) Prove that $1 \in U$.
   b) Prove that U is a subgroup of G.
3. Let G be a linear algebraic group and let $x \in G$. Write $x = su = us$, where u is unipotent and s is semisimple. Using Exercise 2 above, prove that
   a) $u, s \in \overline{\{x^n | n \geq 1\}}$
   b) For any $x \in M_n(k)$ there exists an idempotent in $\overline{\{x^n | n \geq 1\}}$.
4. Let M be an irreducible, algebraic monoid with unit group G, and suppose that $\rho : G' \to G$ is a finite dominant morphism of algebraic groups. Prove that there exists an irreducible algebraic monoid M' with unit group G', and a finite dominant morphism $\psi : M' \to M$, such that $\psi|G' = \rho$.

5. Let G be a connected algebraic group with $X(G) = \{1\}$. Suppose that M is an irreducible, algebraic monoid with unit group G. Prove that $M = G$.
6. Let G be a connected algebraic group with $X(G) \neq \{1\}$.
   a) Let $T \subseteq G$ be a maximal torus and let $1 \neq \chi \in X(G) \subseteq X(T)$ (by restriction). For a finite subset $S \subseteq X(T)$, let $< S > \subseteq X(T)$ denote the submonoid generated by S. Prove that there exists $n > 0$ such that $C =< S + n\chi > \subseteq X(T)$ satisfies $C \cap -C = \{0\}$.
   b) If $\rho : G \to Gl(V)$ is a representation, show that $\overline{\rho(T)}$ has a zero element (of its own) if and only if $C =< \phi_T(V) > \subseteq X(T)$ satisfies $C \cap -C = \{0\}$. Here, $\phi_T(V)$ is the set of weights of T on V via ρ.
   c) Using a), show that there exists a representation $\rho : G \to Gl(V)$ with finite kernel such that the condition of b) is satisfied.
   d) Using c) and Exercise 4 above, show that there exists an irreducible algebraic monoid M with 0 and unit group G.
7. Let M be irreducible and suppose that $x, y \in \overline{T}$. Suppose that there exists $g \in G(M)$ such that $gxg^{-1} = y$. Prove that there exists $h \in N_G(T)$ such that $hxh^{-1} = y$
8. Suppose that $G \subseteq Gl(V)$ is a closed connected subgroup such that V is an irreducible module for G. Prove that
   a) G is reductive.
   b) $\dim Z(G) \leq 1$.
9. Let $\rho : G \to G'$ be a finite dominant morphism of reductive algebraic groups. Prove that $\rho : \mathcal{U} \to \mathcal{U}$ is bijective. Here, $\mathcal{U} = \{g \in G \mid g \text{ is unipotent }\}$.

3.5.2 Linear Algebraic Semigroups

1. Let S be an algebraic semigroup and let $I \subseteq S$ be a two-sided ideal such that $S \backslash I$ is multiplicatively closed. Prove that I is Zariski closed.
2. Let A be a positively graded, finitely generated k-algebra with $A_0 = k$. Let $M = End(A)$, the monoid of degree preserving k-algebra endomorphisms.
   a) Prove that M is an algebraic monoid with 0.
   b) For $e, f \in E(M)$, define $e \sim f$ if $Image(e) \cong Image(f)$ as graded k-algebras. Prove that $E(S)/\sim$ is a finite set.
3. Let A be a k-algebra, and assume that we have a rational action $\rho : k^* \to Aut(A)$. Prove that the following are equivalent.
   a) ρ extends to a rational action $\overline{\rho} : k \to End(A)$.
   b) $A = \oplus_{n \geq 0} A_n$, where $\rho(f)(t) = t^n f$, for all $f \in A_n$.
4. Let $\psi : S \to T$ be a morphism of algebraic semigroups, and suppose that $e \in E(T) \cap \psi(S)$. Prove that $\psi^{-1}(e)$ contains an idempotent.
5. Let M be an algebraic monoid and let $x \in M$. Suppose that x has a *right inverse* $y \in M$ such that $xy = 1$. Prove that y is also a left inverse of x.
6. Let M be an irreducible algebraic monoid, and let $\rho : M \to End(V)$ be an irreducible representation. Let $e \in E(M)$. Prove that $e(V) \subseteq V$ is an irreducible representation of eMe.

7. Let k be a field, and let $E \subseteq M_n(k)$ be an infinite set of idempotents of rank r. Prove there exists $e, f \in E$, $e \neq f$, such that rank(ef)=rank(fe).
8. With E as in Exercise 7 above, but Zariski closed, let $X = \{f \in E | rk(fe) < r \text{ or } rk(ef) < r\}$. Prove that X is a closed, proper subset of X.
9. Suppose that $S \subseteq M_n(k)$ is a closed subsemigroup, and let $e, f \in M_n(k)$ be such that rk(ef)=rk(fe)=rk(e)=rk(f). Show that $e\mathcal{J}f$ in S.
10. Let $\sim$ be the equivalence relation on $E(S)$ generated by $e \sim f$ if rk(ef)=rk(fe)=rk(e)=rk(f). Suppose that $V \subseteq E(S)$ is an irreducible component, and $S \subseteq M_n(k)$ as in Exercise 9. Prove that $e \sim f$ for all $e, f \in V$.
11. Using Exercies 7-10 prove that any connected component of $E(S)$ consists of $\mathcal{J}$-related idempotents. See [111] for more details and some applications to rational homotopy theory. See also [99, 27] for some related applications.

3.5.3 Irreducible Algebraic Semigroups

1. Let S be an irreducible algebraic semigroup, and suppose that $E(S) = S$. Prove that S is a rectangular band (see [15]).
2. Let M be an irreducible algebraic monoid and let $x \in M$. Prove that $\{B \in \mathcal{B} | x \in \overline{B}\}$ is closed in $\mathcal{B}$. Here, $\mathcal{B}$ is the projective variety of Borel subgroups of G.
3. Let M be an irreducible monoid with solvable unit group G, and $0 \in M$. Prove that $\mathcal{N} = \{x \in M | x^n = 0 \text{ for some } n > 0\}$ is a two-sided ideal of M.
4. Let M be irreducible. For $x \in M$, prove that the following are equivalent:
   a) $x \in H_e$, the $\mathcal{H}$-class of some idempotent, and x is a semisimple element in H_e.
   b) For any representation $\rho : M \to End(V)$, $\rho(x)$ is diagonalizable.
5. let M be an algebraic monoid with 0. Prove that the following are equivalent.
   a) M is connected in the Zariski topology.
   b) There exists a chain of idempotents $1 = e_0 > e_1 > \cdots > e_n = 0$ such that $e_i \in \overline{H_{e_{i-1}}}$, for each $i = 1, 2, \ldots, n$.

   See [106].
6. Let M be an algebraic monoid with $M = \overline{G}$. Prove that $E(M) \subseteq \overline{G^0}$.
7. Let M be an irreducible algebraic monoid with 0, such that rk(G)=1. In particular, G is solvable. Prove that every irreducible component of $\mathcal{N} = \{x \in M | x^n = 0, \text{ for some } n > 0\}$ has codimension one in M.
8. Let M be an irreducible monoid with solvable unit group G, and let $X(M) = Hom(M, k)$ be the monoid of characters of M. Let $T \subseteq G$ be a maximal torus, and let $Z = \overline{T}$. Prove that $j^* : X(M) \to X(Z)$ is an isomorphism, where $j : Z \to M$ is the inclusion. Conclude that, for any such M, there exists a D-monoid Z and a morphism $\pi : M \to Z$ such

that $\pi|\overline{T} : \overline{T} \to Z$ is an isomorphism. Furthermore, π is universal for maps from M to D-monoids. (See [101].)

9. Let M be solvable and irreducible with $\pi : M \to Z$ as in Exercise 8 above. An *ideal* of M is a subset $I \subseteq M$ such that $MIM \subseteq I$. We write $I < M$.
   a) Prove that the following are equivalent for $I < M$:
      i) If $x^n \in I$ for some $n > 0$, then $x \in I$.
      ii) $I = \pi^{-1}(\pi(I))$.

      Such ideals are called *radical.* Notice that any radical ideal of M is closed.
   b) Prove that there is one-to-one correspondence between radical ideals of M and ideals of Z. Notice that any ideal of Z is radical.
10. Let M be irreducible and suppose that $I < M$. Let $X = \overline{B} \subseteq M$, where $B \subseteq G$ is a Borel subgroup. Let $I(B) = I \cap M$. Suppose that $I(B)$ is radical. Prove that I is closed in M. Conclude that any prime ideal of M is closed. (An ideal $I < M$ is *prime* if $M \backslash I$ is multiplicatively closed.)
11. Let M be irreducible, and suppose that $P, Q < M$ are prime ideals such that $P \cap \overline{T} = Q \cap \overline{T}$. Prove that $P = Q$. Hint: use exercise 9 above.
12. Let M be an irreducible algebraic monoid, and suppose $\overline{B} \subseteq M$ is the closure of a Borel subgroup of G. Suppose $x \in \overline{B}$ is the zero element of $\overline{B}$. Prove that x is the zero element of M.
13. Let M be an irreducible monoid and let $\eta : M' \to M$ be the normalization of M. Prove that M has the unique structure of an algebraic monoid so that η is a morphism of algebraic monoids.
14. Let M be irreducible. Suppose that $V \subseteq M \backslash G$ is closed, and has codimension one in M. Prove that $V < M$.
15. Let M be reductive, and let $e \in M$. Let $P = \{g \in G | ge = ege\}$. Prove that $R_u(P)e = \{e\}$ (See [95]).
16. a) Let G be a connected, commutative algebraic group in characteristic zero. Prove that, for all $n > 0$, $\psi_n : G \to G$, $\psi_n(x) = x^n$, is a surjective morphism of algebraic groups.
    b) Let M be an irreducible, commutative monoid in characteristic zero.
       i) Prove that $E(M)$ is finite.
       ii) Prove that $\bigcup_{e \in E(M)} H_e$ is a submonoid of M.

4

Regularity Conditions

The major question underlying much of the structure theory of algebraic monoids is the rôle of *idempotents*. Namely, "How is a monoid put together from its unit group and its set of idempotents?"

To identify monoids with interesting structural properties, one needs to assume some kind of *regularity condition* either directly or indirectly. A regularity condition is some structural assumption that allows the idempotents to play a decisive role.

4.1 Reductive Monoids

Definition 4.1. *Let M be an irreducible algebraic monoid.*

a) M is reductive *if $G(M) = \{g \in M \mid g^{-1} \in M\}$ is a reductive group.*
b) M is regular *if $M = G(M)E(M) = E(M)G(M)$*

The next result is due to Putcha and the author. We sketch the proof from Theorems 7.3 of [82].

Theorem 4.2. *Let M be an irreducible monoid with unit group G and zero element $0 \in M$. The following are equivalent.*

a) G is reductive.
b) M is regular.
c) M has no non-zero nilpotent ideals.

Proof. Assume a), and let $S \subseteq M$ be an irreducible component of $M \backslash G$. One then checks that $S = \overline{MeM}$ for some $e \in E(S)$. But $GeMeG$ is closed in M since $B_1 eMeB_2 = eMe$ for appropriate Borel subgroups B_1 and B_2 of G. On the other hand, $GeMeG$ is dense in $S = \overline{MeM}$. Thus $S = GeMeG$. But inductively, eMe is regular, since $G(eMe)$ is an image of the reductive group $C_G(e)$. Thus, S consists of regular elements, and so M is regular.

If M is regular and $a \in M$ then, by definition, $ag = e = e^2$ for some $e \in E(M)$ and $g \in G$. So there can be no non-zero nilpotent ideals.

Assume that G is not reductive, and find a representation $M \subseteq End(V)$ of M as a closed submonoid. Assume that $V_1 \subseteq V_2 \subseteq \cdots \subseteq V_i \subseteq \cdots \subseteq V_n = V$ is an M-stable filtration such that each V_i/V_{i-1} is irreducible over M. Let $M' \subseteq \prod_{i=1}^{n} End\ (V_i/V_{i-1})$ be the semisimplification of M (as in [102]) and notice that we have an induced map

$$\theta : M \longrightarrow M' .$$

Now $ker(\theta|G) = R_u(G)$, and so $\theta^{-1}(0')$ is a nilpotent ideal of M of dimension greater than or equal to $dim(R_u(G))$.

But what happens if M does not have a zero element? The following answer is due to Putcha (see Theorem 7.4 of [82]). For $e \in E(M)$, let

$$G_e = \{g \in G \mid ge = eg = e\ \}.$$

Recall that a monoid M is *completely regular* if $M = \bigsqcup_{e\in E(M)} H_e$. For example, any commutative, regular monoid is completely regular.

Theorem 4.3. *Suppose that M is irreducible with unit group G and minimal idempotent $e \in E(M)$. Then the following are equivalent.*

a) M is regular.
b) $\overline{R(G)}$ is completely regular.
c) $G_e := \{g \in G \mid ge = eg = e\}^0$ is reductive.

Proof. If M is regular and $a \in \overline{R(G)}$ then $axa = a$ for some $x \in M$. But $M = \cup \overline{B}$ where B is a Borel subgroup of G. Hence $x \in \overline{B}$ for some B. But also, $a \in \overline{R(G)} \subseteq \overline{B}$. Thus $a \in \overline{B}$ is a regular element of $\overline{B}$. But then $a\mathcal{H}e$ for some $e \in E(\overline{B})$ by 3.19 and 4.12 of [82]. Consequently $e \in \overline{R(G)}$ by an easy calculation.

If $\overline{R(G)}$ is completely regular, first notice that $e \in \overline{R(G)}$, since all minimal idempotents are conjugate. One then checks that $R(G_e) \subseteq R(G)_e$, while the latter group is a torus. Thus G_e is reductive.

If G_e is reductive then, by Theorem 4.2, $M_e := \overline{G_e}$ is regular. But from 6.13 of [82] $M = GM_eG$, since $e \in K = ker(M)$, the minimal, (regular) $\mathcal{J}$-class of M.

Further results have been obtained by Huang in [39]. In particular, he obtains the following characterization of reductive monoids.

Theorem 4.4. *Let M be an irreducible algebraic monoid. Then the following are equivalent.*

a) M is reductive.
b) M is regular and the semigroup kernel of M is a reductive group.

A similar result was obtained by Rittatore in [121].

4.2 Semigroup Structure of Reductive Monoids

Reductive monoids enjoy the richest and most interesting structural properties. Let M be reductive with unit group G and Borel subgroup $B \subseteq G$ with maximal torus $T \subseteq B$. Let $E = E(M)$, and let P, Q denote parabolic subgroups of G. $Cl_G(e)$ is the conjugacy class of e in M. The following result was first observed by Putcha in [85].

Theorem 4.5. *Let M be an irreducible algebraic monoid with reductive unit group G.*

a) $P(e) = \{x \in G \mid xe = exe\}$ *and* $P^-(e) = \{x \in G \mid ex = exe\}$ *are opposite parabolic subgroups such that* $eR_u(P^-) = R_u(P)e = \{e\}$.
b) *If* $e, f \in E$ *with* $eM = fM$ *or* $Me = Mf$, *then* $x^{-1}ex = f$ *for some* $x \in G$.
c) $\Lambda := \{e \in E(\overline{T}) \mid Be = eBe\} \cong G\backslash M/G$. *In particular,*

$$M = \bigsqcup_{e \in \Lambda} GeG$$

d) $Cl_G(e) \cong \{(P, Q) \mid P$ *and* Q *are opposite and there exists* $x \in G$ *such that* $x^{-1}Px = P(e)\}$.

4.2.1 The Type Map

Definition 4.6. *a)* Λ *is the* cross section lattice *of M relative to T and B.*
b) The type map $\lambda : \Lambda \to 2^S$ *is defined as follows:* $\lambda(e) \subseteq S$ *is the unique subset such that* $P(e) = P_{\lambda(e)}$.

The type map is the most important single combinatorial invariant in the structure theory of reductive monoids. It is in some sense, the monoid analogue of the Coxeter-Dynkin graph. It also shows us which $G \times G$-orbits are involved in the monoid, as well as how the monoid structure is built up from these orbits [85, 86]. See § 10.4 for more discussion on how this is done.

Example 4.7. The type map of $M_{n+1}(K)$. Let $S = \{s_1, \ldots, s_n\}$ be the simple involutions of $Sl_n(K)$ ordered in the usual way. Let

$$e_i = \begin{pmatrix} 1 & & & & & \\ & \ddots & & & & \\ & & 1 & & & \\ & & & 0 & & \\ & & & & \ddots & \\ & & & & & 0 \end{pmatrix} \qquad i = 1, \ldots, n+1\,;$$

Then $\Lambda = \{0, e_1, \ldots, e_{n+1}\}$ and, for $i \geq 2$,

$$\lambda(e_i) = \{s_1, \ldots, s_{i-1}\} \cup \{s_{i+1}, \ldots, s_n\}$$

See Chapter 7 for more sample calculations of the type map along with the corresponding lattice diagrams.

Putcha has used Theorem 4.5 as a basis for the "abstract" theory. Indeed, let G be a group with BN pair and finite Weyl group. He then defines a **monoid of Lie type** to be an abstract monoid M with unit group G satisfying a) and b) above, as well as being generated by G and $E(M)$. Properties c) and d) then follow automatically. Such monoids are classified up to central extension by the type map $\lambda : \Lambda \to \mathcal{P}$, $\lambda(e) = P(e)$ where $\mathcal{P} = \{P < G \mid B \subseteq P\} \cong 2^S$. See Chapter 10 for the main details of this surprising development. For example, to see how one constructs a monoid of Lie type from a type map $\lambda : \Lambda \to 2^S$.

In § 5.3.3 the type map is described in terms of the standard classification mechanism of spherical embeddings.

Theorem 4.8. *Let M be an reductive algebraic monoid with unit group G, and cross section lattice Λ. Let $e \in \Lambda$.*

a) Define

$$eMe = \{x \in M \mid x = exe\}.$$

Then eMe is a reductive algebraic monoid with unit group $H_e = eC_G(e)$. A cross section lattice of eMe is $e\Lambda = \{f \in \Lambda \mid fe = f\}$.

b) Define

$$M_e = \overline{\{x \in G \mid ex = xe = e\}^0}.$$

Then M_e is a reductive monoid with zero $e \in M$ and unit group $\{x \in G \mid ex = xe = e\}^0$. A cross section lattice for M_e is $\Lambda_e = \{f \in \Lambda \mid fe = e\}$.

The reader is refered to [82] for the proof.

4.3 Solvable Regular Monoids

Let M be an irreducible normal monid with solvable unit group G. The assumption that M is regular imposes decisive restrictions on the structure of M. The results of this section were first recorded in [108]. Write

$$G = TU = UT,$$

where U is the unipotent radical of G, and $T \in G$ is a maximal torus of G. Consider $X(\overline{T}) \subseteq X(T)$, the set of characters of $\overline{T}$. As indicated, $X(\overline{T})$ can be thought of as the set of characters of T that extend over $\overline{T}$. Let $e \in E(\overline{T})$ be the minimal idempotent. We assume, with little loss of generality, that e is the zero element of $\overline{T}$. Let

$$\begin{aligned} U_+ &= \{u \in U \mid eu = e\} \\ U_0 &= \{u \in U \mid eu = ue\} \\ U_- &= \{u \in U \mid ue = e\}. \end{aligned}$$

Theorem 4.9. *a)* $m : U_+ \times U_0 \times U_- \longrightarrow U$ *is an isomorphism.*
b) $U_0 = C_U(T)$
c) $T \times U_+ \longrightarrow U_+,\ (t,u) \longmapsto tut^{-1}$ *extends to an action of* $\overline{T}$ *on* U_+.
d) $T \times U_- \longrightarrow U_-,\ (t,u) \longmapsto t^{-1}ut$ *extends to an action of* $\overline{T}$ *on* U_-.
e) $m : U_+ \times M_0 \times U_- \longrightarrow M$ *is an isomorphism of varieties, where* $M_0 = \overline{TU_0} \cong \overline{T} \times U_0$.
f) $\Phi_T(U) = \{\alpha \in X(T) \mid \mathfrak{L}(U)_\alpha \neq 0\}$ *is contained in* $X(\overline{T}) \cup -X(\overline{T})$.

A special case of this was first considered on page 182 of [47], and some of those embeddings were observed to be algebraic monoids.

The multiplication law on M can be defined in terms of the coordinates of Theorem 4.9 e). First let

$$U = U_+U_0U_- \cong U_+ \times U_0 \times U_- \ .$$

Then $U_-U_+ \subseteq U$ defines

$$\begin{aligned} \zeta_+ &: U_- \times U_+ \longrightarrow U_+ \\ \zeta_0 &: U_- \times U_+ \longrightarrow U_0 \\ \zeta_- &: U_- \times U_+ \longrightarrow U_- \end{aligned}$$

so that for $u \in U_-$ and $v \in U_+$

$$uv = \zeta_+(u,v)\zeta_0(u,v)\zeta_-(u,v) \ .$$

Now recall from Theorem 4.9 c) and d) above, the actions

$$\begin{aligned} a_+ &: \overline{T} \times U_+ \longrightarrow U_+ \\ a_- &: \overline{T} \times U_- \longrightarrow U_- \end{aligned}$$

extending the action of T by inner automorphisms. Denote

$$a_+(x,u) \text{ by } u^{\overline{x}}$$

and

$$a_-(y,u) \text{ by } u^{\overline{y}}.$$

From formula (4) of [108], we obtain the following multiplication table.

Proposition 4.10. *The morphism of 4.9 e) is an isomorphism of algebraic monoids if we define the product on* $U_+ \times M_0 \times U_-$ *by*

$$(u,x,v)(a,y,b) = (u\zeta_+(v,a)^{x}, x\zeta_0(v,a)y, \zeta_-(v,a)^{\overline{y}}b) \ .$$

There are converses to Theorem 4.9 and Proposition 4.10.

Proposition 4.11. *Let* U *be a connected, unipotent algebraic group and suppose we are given the following data.*

a) *A torus action* $\rho : T \longrightarrow Aut(U)$ *by algebraic group automorphisms.*
b) *A normal torus embedding* $T \subseteq \overline{T}$ *such that* $0 \in \overline{T}$ *and* $\Phi_T(U) \subseteq X(\overline{T}) \cup -X(\overline{T})$.

Then there exists a unique structure of a regular algebraic monoid on $M = U_+ \times \overline{T} \times C_U(T) \times U_-$ *extending the group law on* $U_+ \times T \times C_U(T) \times U_- = G = UT$.

Proof. Since $\Phi_T(U) \subseteq X(\overline{T}) \cup -X(\overline{T})$, the actions $T \times U_+ \longrightarrow U_+$ and $T \times U_- \longrightarrow U_-$ both extend over $\overline{T}$. Thus we can define a monoid structure on M using the formula

$$(u, x, v)(a, y, b) = (u\zeta_+(v, a)^x, x\zeta_0(v, a)y, \zeta_-(v, a)^{\overline{y}}b) .$$

of 4.10.

4.4 Regular Algebraic Monoids

In this section we deal with the general case: the structure of any normal, regular algebraic monoid M with arbitrary (connected) unit group G. These results were first recorded in [117]. As in [117], we proceed in two steps. This makes things easier to understand.

So let M be a normal, regular monoid with unit group G. By Theorem 4.3, G_e is reductive for any minimal idempotent e of M.
Step 1: Assume that G_e is a Levi factor of G. Thus G is the semidirect product

$$G = G_e \ltimes U$$

of G_e and U, where $U = R_u(G) \lhd G$ is the unipotent radical of G.

Theorem 4.12. *Let* $T \subseteq G$ *be a maximal torus and let* $\overline{T} \subseteq M$ *be the Zariski closure of* T *in* M. *Let* $\Phi_T(U) \subseteq X(T)$ *be the weights of the action* $Ad : T \longrightarrow Aut(\mathfrak{L}(U))$ *on the Lie algebra of* U. *Then* $\Phi_T(U) \subseteq X(\overline{T}) \cup -X(\overline{T})$.

Conversely, suppose that we are given an algebraic group G *of the form* $G = G_0 \ltimes U$, *where* $G_0 \subseteq G$ *is a Levi factor, along with a normal algebraic monoid* M_0 *with zero and unit group* G_0 *and maximal* D*-submonoid* $\overline{T} \subseteq M_0$. *Consider the action* $Ad : T \longrightarrow Aut(\mathfrak{L}(U))$ *and assume that* $\Phi_T(U) \subseteq X(\overline{T})$. *Then there exists a unique, normal, algebraic monoid* M *with unit group* G *and maximal* D*-submonoid* $\overline{T} \subseteq M$.

Proof. Let $N = \overline{TU} \subseteq M$ and let $e \in E(\overline{T})$ be a minimal idempotent of M. Since G_e is a Levi factorof G, any maximal torus of $C_G(e)$ is in G_e. Then e, being the zero element of $\overline{G_e}$, is the zero element of $\overline{T}$. This is precisely the point of the assumption in “Step 1”.

Then $(TU)_e \cap U \subseteq G_e \cap U = \{1\}$. Hence $(TU)_e$ is a connected solvable group with no unipotent elements other than 1. So $(TU)_e$ is a torus, and thus, by 4.3, N is regular. Hence, by Theorem 4.9 f), $\Phi_T(U) \subseteq X(\overline{T}) \cup -X(\overline{T})$.

Conversely, assume that we have $\rho : G_0 \longrightarrow Aut(U)$ such that $\Phi_T(U) \subseteq X(\overline{T}) \cup -X(\overline{T})$. Then we can write

$$U = U_+ U_0 U_-$$

where

$$\mathfrak{L}(U_+) = \bigoplus_{\alpha \in X(\overline{T})} \mathfrak{L}(U)_\alpha$$
$$\mathfrak{L}(U_0) = C_{\mathfrak{L}(U)}(T)$$

and

$$\mathfrak{L}(U_-) = \bigoplus_{-\alpha \in x(\overline{T})} \mathfrak{L}(U)_\alpha .$$

One checks that U_+, U_0 and U_- are normalized by G_0 since we can define them (as above) using a central one parameter subgroup of G_0. Therefore define M by

$$M = U_+ \times M_0 \times U_0 \times U_- \ .$$

Now the action $\rho : G_0 \longrightarrow Aut(U_+)$ extends to $\rho : M_0 \longrightarrow End(U_+)$ and $\rho^{-1} : G_0 \longrightarrow Aut(U_-)$ extends to $\rho^{-1} : M_0 \longrightarrow End(U_-)$ (using the **opposite monoid** M_0^{op} in the latter case).

Hence we can define the multiplication on M using the formula of Proposition 4.10.

Step 2: Now let M be any normal, irreducible, regular algebraic monoid with unit group G. Let $e \in E(M)$ be a minimal idempotent. Let

$$H = G_e R_u(G) < G \ ,$$

and define

$$N = \overline{H} \subseteq M \ .$$

Theorem 4.13. *a) N is a regular monoid of the type considered in Theorem 4.12. Furthermore, $gNg^{-1} = N$ for all $g \in G$.*

b) Define $N \times^H G = \{[x,g] \mid x \in N, g \in G\}$ where $[x,g] = [y,b]$ if there exists $k \in H$ such that $y = xk^{-1}$ and $h = kg$. Then $N \times^H G$ is a regular algebraic monoid with multiplication $[x,g][y,h] = [xgyg^{-1}, gh]$. Furthermore,

$$\varphi : N \times^H G \longrightarrow M$$
$$\varphi([x,g]) = xg$$

is an isomorphism of algebraic monoids.

Proof. Let $g \in G$. Then $gG_eg^{-1} = G_{geg^{-1}}$. But $geg^{-1} = heh^{-1}$ for some $h \in G_eR_u(G)$ by Theorem 6.30 of [82]. Thus $gG_eg^{-1} = hG_eh^{-1}$ and so $gG_eR_u(G)g^{-1} = G_eR_u(G)$. Hence $gNg^{-1} = N$. To see that N is regular, notice that $G_eR_u(G))_e = G_e$ and so by Theorem 4.3 N is regular.

Now $G_e \times R_u(G) \longrightarrow G_eR_u(G)$ is bijective, and its kernel is $G_e \cap R_u(G)$, which is an infinitesimal unipotent group scheme. One checks, as in Lemma 4.1 of [117], that $G_e \cap R_u(G)$ is actually central in G_e, yet $Z(G_e)$ is a diagonalizable group scheme. Thus, N is a regular monoid of the type considered in Theorem 4.12.

For b) one checks that φ is surjective and birational while M is normal. Thus, φ is an isomorphism.

4.5 Regularity in Codimension One

It follows from Theorem 4.2 above that any normal, reductive monoid M is determined by the diagram

$$\overline{T} \supseteq T \subseteq G\,.$$

Hence, if we know the closure of a maximal torus in M, we can identify M to within isomorphism. Furthermore, any $\overline{T} \supseteq T \subseteq G$, as above, for which

a) the Weyl group action on T extends over $\overline{T}$,
b) $\overline{T}$ is a normal affine variety,

occurs for some M.

The main reason for this rigidity in the classification is the *Extension Principle* (§ 5.1) enjoyed by all normal reductive monoids and which we now state.

Any morphism $\alpha : G \longrightarrow N$ of algebraic monoids which extends over $\overline{T}$ via

$$\begin{array}{ccc} T & \xrightarrow{\alpha|T} & N \\ \cap & \nearrow & \\ \overline{T} & & \end{array}$$

can be extended to a unique morphism $\beta : M \longrightarrow N$ of algebraic monoids. This extension property (EP) results largely from the fact that reductive monoids are *regular*. In fact, $M = \bigcup_{e \in \Lambda} GeG$, a condition that should make one suspect that M has the EP.

So what about nonreductive monoids? It is easy to see from simple examples that

(1) Not every M is regular.
(2) Not every M satisfies EP.

(3) There exist morphisms (unlike the case of reductive monoids) of nonreductive monoids $\varphi : M' \longrightarrow M$ such that
$\varphi : \overline{T}' \underset{\cong}{\longrightarrow} \overline{T}$,
$\varphi : G' \underset{\cong}{\longrightarrow} G$ and
φ is not a finite-to-one morphism.

To illustrate (3), define

$$M' = \{(s,t,u) \mid s,t,u \in k\}$$

with

$$(s,t,u)(k,\ell,v) = (sk, t\ell, \ell u + sv)$$

and

$$M = \{(s,t,u) \mid s,t,u \in k\}$$

with

$$(s,t,u)(k,\ell,v) = (sk, t\ell, k\ell u + s^2 v).$$

Finally, define $\varphi : M' \longrightarrow M$ by

$$\varphi(s,t,u) = (s,t,su)\ .$$

So there are some very significant differences. However, there are some very compelling open questions here, and they are all related.

Problem 4.14. Given an irreducible monoid M_1, does there exist an irreducible monoid M and a morphism $\varphi : M \longrightarrow M_1$ such that
a) $\varphi : G \underset{\cong}{\longrightarrow} G_1$,
b) $\varphi : \overline{T} \underset{\cong}{\longrightarrow} \overline{T}_1$ and
c) M satisfies EP relative to $\overline{T} \supseteq T \subseteq G$?

Problem 4.15. Given $\overline{T} \supseteq T \subseteq G$ so that the W action on T extends over $\overline{T}$, does there exist an M realizing these data?

Problem 4.16. Are the following equivalent for M normal?
a)M has the EP relative to $\overline{T} \supseteq T \subseteq G$.
b)M is regular in codimension one.

We say an irreducible monoid M is *regular in codimension one* if there exists a closed two-sided ideal $I \subseteq M$ such that

(i) $dim(I) \leq dim(M) - 2$
(ii) $M \backslash I \subseteq \bigcup_{e \in E(M)} GeG$.

Problem 4.17. If M has the EP, is M regular in codimension one?

Remark 4.18. Notice that it is not always possible to find a regular monoid with given $\overline{T} \supseteq T \subseteq G$. (See [103] for the precise restrictions when G is solvable.) For example, no regular monoid shares $D_n(k) \supseteq D_n(k)^* \subseteq T_n(k)^*$ with $T_n(k)$. It is hoped that regularity in codimension one is the exact general condition that allows us to extend known results about reductive monoids to the general case.

4.6 Exercises

4.6.1 D-monoids

1. Assume that M is reductive and let $x \in M$. Prove that $\dim(Cl_G(x)) \leq \dim M - \dim T$.
2. Let $Z = \{(a,b,c,d) \in K^4 | ab = cd\}$. Find the lattice of idempotents of Z. What is the polytope involved?
3. Let M be irreducible and consider $X(M) \subseteq K[M]$, the monoid of characters of M. Prove that $X(M)$ is a linearly independent subset of $K[M]$.
4. Let Z be a D-monoid with zero, and assume that $\sigma : Z \to Z$ is an automorphism of Z such that $\sigma(e) = e$ for any $e \in E(Z)$. Prove that σ is the identity automorphism. Conclude that $Aut(Z)$ is a finite group.
5. Let Z be a D-monoid with zero, and let $E^1(Z) = \{e \in E(Z) | 1 \text{ covers } e\}$.
   a) Show that $G_e = K^*$.
   b) Prove that, for each $e \in E^1(Z)$, there exists an unique injective morphism $\alpha_e : K \to Z$ such that $\alpha(K^*) = G_e$ and $\alpha(0) = e$.
   c) Consider the morphism $\psi : K \times \cdots \times K \to Z$ defined by $\psi(t_1, \ldots, t_s) = \alpha_{e_1}(t_1) \ldots \alpha_{e_s}(t_s)$, where $E^1(Z) = \{e_1, \ldots, e_s\}$. Prove that ψ is surjective, and that ψ induces a bijection on E^1.
   d) Using c), prove that the following are equivalent:
      i) ψ is a finite morphism.
      ii) $E(Z)$ is a Boolean lattice.
      iii) For any $e, f \in E^1(Z)$, $e \neq f$, $ef \in E^2(Z)$.
6. Let Z be a one-dimensional, normal D-monoid with zero. Prove that $Z \cong K$.
7. Let Z be a normal D-monoid with zero. For $e \in E_1(Z)$ define $\chi_e : Z \to eZ \cong K$ by $\chi_e(z) = ez$. Define $\psi : Z \to K^m$, $m = |E_1(Z)|$, by $\psi(z) = (\chi_e(z))_{e \in E_1(Z)}$. Prove that ψ is a finite morphism.
8. Let M be reductive and let $x \in M$. Assume that $x \in \overline{T}$. Prove that $Cl_G(x) \subseteq M$ is closed.
9. Let Z be a three dimensional D-monoid with zero. Prove that $|E_1(Z)| = |E^1(Z)|$. Prove that any $n \geq 3$ can occur as $|E_1(Z)|$.
10. Let Z be a D-monoid and let $e \in E(Z)$. Prove that $Z_e = \{x \in Z | ex\mathcal{H}e\}$ is the unique open submonoid of Z with e as minimal idempotent. Prove that there exists a morphism $\chi : Z \to K$ such that $Z_e = \{z \in Z \mid \chi(z) \neq 0\}$.

4.6.2 Regular and Reductive monoids

1. Let M be regular, and assume that $I \subseteq M$ is a two-sided ideal of M. Prove that $I = \overline{I}$.
2. Let $\rho : M \to N$ be a morphism of algebraic monoids such that
   a) $\rho|\overline{T}$ is bijective,
   b) $\rho(G(M)) = G(N)$,
   c) N is regular.

Prove that $\rho(M) = N$.

3. Prove that for any irreducible monoid M there exists an irreducible monoid N and a morphism $\rho : M \to N$ satisfting the three conditions of Exercise 2 above.
4. Let M be reductive with zero. Suppose that $\sigma : M \to M$ is an automorphim such that $\sigma(x)\mathcal{J}x$ for any $x \in M$. Prove that $\sigma(x) = gxg^{-1}$ for some $g \in G$.
5. Let M be reductive with zero and dim$Z(G)$=1. Assume also that (G, G) is simple. Prove that there exist morphisms $\rho_i : M \to M_i, i = 1, \ldots, s$, such that
   a) Each $\rho_i : G \to G_i$ is an isomorphism.
   b) Each M_i has a unique, nonzero, minimal $\mathcal{J}$-class.
   c) $\rho : M \to M_1 \times \cdots \times M_s$, $\rho(x) = (\rho_1(x), \ldots, \rho_s(x))$, induces a bijection on minimal nonzero $\mathcal{J}$-classes.
6. Let M be reductive and let $x \in M$. Prove there exists a maximal torus T, an idempotent $e \in E(\overline{T})$, and $\sigma \in N_G(T)$ such that $x\mathcal{H}e\sigma$.
7. Let M be irreducible and reductive. Prove there exists a unique, minimal idempotent of M which is central.
8. Let M be irreducible and normal. Prove that $\overline{T}$ is also normal.

4.6.3 Regularity in Codimension One

1. Let A be a finite-dimensional K-algebra. Prove that the following are equivalent.
   a) A is regular in codimension one, as an algebraic monoid.
   b) For any primitive idempotent $e \in E(A)$, dim(eAe)=1.
2. Let M be reductive, and let P be a parabolic subgroup of $G(M)$. Prove that $\overline{P}$ is regular in codimension one.

5

Classification of Reductive Monoids

Associated with any reductive monoid M, are the unit group G of M, and the maximal D-submonoid $Z = \overline{T} \subseteq M$ of M. Here, $T \subseteq G$ is a maximal torus of G. The classification of reductive groups follows the well-established identification in terms of Dynkin diagrams and root systems [7]. On the other hand Z is an affine torus embedding of T and, as such, it is classified by its **integral polyhedral cone** $X(Z) \subseteq X(T)$. $X(Z)$ is just the set of characters of Z, as defined in the previous chapter. Integral polyhedral cones were first discussed in [104], in the classification problem. G and Z are tied together by the Weyl group $W = N_G(T)/T$. The natural action of W on T extends to an action of W on Z.

Our approach to the classification of these monoids is straightforward. Assume that M is normal and reductive. As above, we have the diagram,

$$Z \supseteq T \subseteq G$$

where Z is a normal T-embedding with W-action extending the natural W-action on T. We explain in this chapter how M is uniquely determined by this diagram.

Conversely, given such a diagram we show how to construct the monoid M so that the Zariski closure $\overline{T} \subseteq M$ of T in M is T-isomorphic to Z. The main ideas behind § 5.1 are taken from [104].

5.1 The Extension Principle

Let M be a normal, reductive monoid with unit group G and Borel subgroup $B \subseteq G$. Let $T \subseteq B$ be a maximal torus. Recall from Definition 4.6 that

$$\Lambda = \{e \in E(\overline{T}) \mid Be = eBe\} ,$$

the cross-section lattice of M associated with B and T. For example, let $M = M_n(K)$, $B = T_n(K)$, $T = D_n(K)$. Then $\Lambda = \{e_o, e_1, \ldots, e_n\}$ where e_i is the rank $= i$ matrix

$$\begin{pmatrix} 1 & & & & & \\ & \ddots & & & 0 & \\ & & 1 & & & \\ & 0 & & 0 & & \\ & & & & \ddots & \\ & & & & & 0 \end{pmatrix}.$$

The following result provides an analogue of the **big cell** for algebraic monoids.

Lemma 5.1. *Let $B^- \subseteq G$ be the Borel subgroup opposite of B relative to T. Let $e \in \Lambda$ be such that $\dim(eT) = \dim(T) - 1$. Let*

$$\begin{aligned} Z_e &= \{x \in \overline{T} \mid ex \in eT\} \\ &= T \cup eT\,. \end{aligned}$$

Define $m : U^- \times Z_e \times U \longrightarrow M$ by $m(x,y,z) = xyz$. Then m is an open embedding.

Proof. Notice first from Corollary 4.3 of [104] that $U^- \longrightarrow U^-e$, $u \mapsto ue$, is bijective; and similarly for $U \longrightarrow eU$, $v \mapsto ev$.

Now suppose that $u_1x_1v_1 = u_2x_2v_2$, where $u_i \in U^-$, $v_i \in U$ and $x_i \in Z_e$. Then $u_2^{-1}u_1x_1 = x_2v_2v_1^{-1}$. But $u_2^{-1}u_1x_1 \in \overline{B^-}$ while $x_2v_2v_1^{-1} \in \overline{B}$. However, $\overline{B} \cap \overline{B^-} = \overline{T}$, using Proposition 27.2 of [40], since in any representation ρ of G we can put $\rho(B^+)$ and $\rho(B^-)$ in the upper and lower triangular form, respectively, by choosing a suitable basis of weight vectors. Thus $u_2^{-1}u_1x_1 = x_2v_2v_1^{-1} \in \overline{T}$. It then follows easily that $u_2^{-1}u_1 = 1$ and $v_2v_1^{-1} = 1$, and so $x_1 = x_2$ as well. This proves that

$$m : U^- \times Z_e \times U \longrightarrow M$$

is injective. But from Proposition 28.5 of [40] we know that m is also birational. Hence, by Theorem 2.29, it is actually an open embedding since M is normal.

The following extension principle was first recorded in Corollary 4.5 of [104].

Theorem 5.2. (Extension Principle) *Let M be normal and reductive, and suppose M' is any algebraic monoid. Suppose that $\alpha : G \longrightarrow M'$ and $\beta : \overline{T} \to M'$ are morphisms of algebraic monoids with $\alpha|T = \beta|T$. Then there exists a unique morphism $\varphi : M \to M'$ such that $\varphi|G = \alpha$ and $\varphi|\overline{T} = \beta$.*

Proof. By the codimension two condition Theorem 2.26, it suffices to extend α to a morphism $\gamma : U \to M'$ where $U \subseteq M$ is any Zariski open set with $codim_M(M\backslash U) \geq 2$.

Therefore we let V be an irreducible component of $M\backslash G$. By Thereom 4.5, there exists $e \in \Lambda \cap V$ such that $\dim(eT) = \dim(T) - 1$. But then from Lemma 5.1, $U_e \cong U^- \times Z_e \times U$ is an open subset of M such that $U_e \cap V \neq \phi$.

Thus, define $U = \left(\bigcup_{e\in\Lambda^1} U_e\right) \cup G$. Here, $\Lambda^1 \subseteq \Lambda$ is the subset of maximal idempotents of $\Lambda\backslash\{1\}$. It suffices to define $\varphi_e : U_e \longrightarrow M'$ by

$$\varphi_e(u,x,v) = \alpha(u)\beta(x)\alpha(v) \ .$$

Clearly the φ_e's patch together to yield the desired morphism $\gamma : U \to M'$.

Remark 5.3. Theorem 5.2 has an appealing interpretation in terms of weights. Let M be a normal, reductive monoid with diagram $\overline{T} \supseteq T \subseteq G$. Suppose that $\rho : G \longrightarrow G\ell(V)$ is a rational representation of G. Then the following are equivalent.

a) ρ extends over M to a representation $\overline{\rho} : M \longrightarrow End(V)$.
b) The set of weights of $\rho|T$, say $\Phi_T(\rho)$, is contained in $X(\overline{T})$.

As a simple example, consider the classical adjoint, $\rho : G\ell_n(K) \longrightarrow G\ell_n(K)$, defined by $\rho(A) = Adj(A)$. Now $G\ell_n(K)$ is the unit group of $M_n(K)$. If $T \subseteq G\ell_n(K)$ is the maximal torus of diagonal matrices, then $\overline{T} = D_n(K) \subseteq M_n(K)$. Thus $X(\overline{T}) = \langle \chi_1, \ldots, \chi_n\rangle$, where $\chi_i(t_1,\ldots,t_n) = t_i$ (the i-th projection). By the above criterion, ρ extends to $\overline{\rho} : M_n(K) \longrightarrow M_n(K)$ since

$$\Phi_T(\rho) = \{\chi_1 \cdot \cdots \cdot \widehat{\chi_i} \cdot \cdots \cdot \chi_n \mid i = 1,\ldots,n\} \subseteq X(\overline{T}) = \langle \chi_1,\ldots,\chi_n\rangle.$$

Of course, the perceptive reader already knows that the adjoint can be defined directly on all of $M_n(K)$ without appealing to weights.

Returning now to our classification problem, we state our main theorem.

Theorem 5.4. *a) Let G be a reductive group with maximal torus $i : T \subseteq G$ and let $j : T \subseteq Z$ be a normal affine torus embedding such that the Weyl group action on T extends over Z. Then there exists a normal monoid M such that $\overline{T} \subseteq M$ is isomorphic to Z.*
b) Let M be any normal reductive monoid with unit group G and maximal D-submonoid $Z = \overline{T} \subseteq M$. Then Z is normal.
c) If M_1 and M_2 are such that $Z_1 \supseteq T_1 \subseteq G_1$ and $Z_2 \supseteq T_2 \subseteq G_2$ are isomorphic as diagrams of the algebraic monoids. Then this isomorphism extends to an isomorphism $M_1 \cong M_2$.

Proof. a) $j^* : X(Z) \to X(T)$ identifies $X(Z)$ with a "convex" subset of $X(T)$. By this, we mean that $X(Z)$ is the intersection of a convex subset of $X(T)\otimes\mathbb{R}$ with $X(T)$. For each dominant $\lambda \in X(Z)$ there exists an irreducible representation (ρ_λ, V) of G such that λ is the highest weight of ρ_λ. It follows from Proposition 3.5 of [104] that the weights of ρ_λ, say $\Phi(\rho_\lambda)$, are contained in $X(Z)$ (*proof*: $\Phi(\rho_\lambda) \subseteq Conv(W \cdot \lambda)$, since λ is the highest weight. But $X(Z)$ is convex). Thus, $\rho_\lambda : G \longrightarrow G\ell(V_\lambda)$ has the property that $\rho_\lambda|T$ extends to $\rho_\lambda : Z \longrightarrow End(V_\lambda)$. Thus we choose $\{(\rho_{\lambda_i}, V_{\lambda_i})\}_{i=1}^{s}$, as above, such that

$\bigcup_{i=1}^{s} W \cdot \lambda_i \subseteq X(Z)$ generates this monoid. Define $M_1 = \overline{\rho(G)} \subseteq End(V)$ where $\rho = \oplus_{i=1}^{s} \rho_{\lambda_i}$ and $V = \oplus_{i=1}^{s} V_{\lambda_i}$. It follows easily that M_1 is a reductive monoid with unit group G, such that $T \subseteq \overline{T}$ is T-isomorphic to $T \subseteq Z$. However, M_1 may not be normal. Hence let $M = \widetilde{M_1}$, the normalization of M_1.

b) Let M be normal, and consider $Z \supset T \subset G$ where Z is the closure of T in M. The issue here is the normality of Z. Hence let $\alpha : \widetilde{Z} \to Z$ be the normalization. But then $\widetilde{Z} \supseteq T \subseteq G$ is as in a). Thus there exists a normal monoid $\widetilde{M}$ with this $\widetilde{Z} \supset T \subset G$. But now we have a morphism of diagrams

$$\begin{array}{ccccc} \widetilde{Z} & \supseteq & T & \subseteq & G \\ \alpha\downarrow & & \delta\downarrow & & \downarrow\beta \\ Z & \supseteq & T & \subseteq & G \end{array}$$

where $\delta = id_T$, $\beta = id_G$.

But by Theorem 5.2 we obtain a unique morphism $\varphi : \widetilde{M} \to M$ such that $\varphi|\widetilde{Z} = \alpha$ and $\varphi|G = \beta$. One checks that φ is finite and birational. But M is normal, and so φ is an isomorphism by Zariski's main theorem. In particular, $\alpha : \widetilde{Z} \to Z$ is an isomorphism, and so Z is normal.

c) follows directly from Theorem 5.2.

Theorem 5.2 has many interesting and useful consequenses. In Chapter 9 it is used to obtain a concise understanding of the finite dimensional representations of reductive normal algebraic monoid. In [104] we obtained the following theorem.

Theorem 5.5. *Let M be a reductive algebraic monoid with one-dimensional center and zero element $0 \in M$. If M is smooth as an algebraic variety, then $M \cong M_n(K)$ as algebraic monoids.*

Proof. We sketch the basic idea of the proof. We refer the reader to [104] for the details. Let $T \subseteq G(M)$ be a maximal torus of $G(M)$. By Theorem 5.2 above, one is essentially reduced to showing that $G(M) \cong Gl_n(K)$ in such a way that $\overline{T}$ is isomorphic to $D_n(K)$.

First notice that $Z = \overline{T} = \{x \in M \mid xt = tx \text{ for all } t \in T\}$. For suppose that $x \in \{x \in M \mid xt = tx \text{ for all } t \in T\}$. Then x is a semisimple element of M, in the sense that its G-conjugacy class is closed in M. Also, x is in the closure of some Borel subgroup B of G. The set of these subgroups can be regarded as a closed subset of G/B. By the Borel Bixed Point Theorem (Theorem 2.36), we may assume that $T \subseteq B$ since T is solvable. From linear algebra and the Lie-Kolchin theorem we may assume that $\overline{B} \subseteq T_n(K)$ (upper-triangular monoid) for some n, and that $\{x\} \cup \overline{T} \subseteq D_n(K)$ (diagonal monoid). The composite

$$\{x\} \cup \overline{T} \subseteq D_n(K) \subseteq T_n(K) \to D_n(K)$$

is injective, where $T_n(K) \to D_n(K)$ is the projection to the diagonal. But the image of $\overline{B}$ and the image of $\overline{T}$ are the same in $D_n(K)$. Thus $x \in \overline{T}$.

Hence Z is the centralizer of T acting on the smooth variety M. By well known results (see for example [42]) Z is a smooth variety. But Z is an affine torus embedding with zero. It follows easily that $Z \cong K^n$. Now $Aut(K^n) \cong S_n$, and so the Weyl group W of T, is a subgroup of S_n. However, it must be generated by reflections, and these are easily identified. In fact, if $K[Z] = K[\chi_1, \ldots, \chi_n]$, then the reflections of W are among those of S_n. Each of these is of the form $\sigma_{i,j}$, which interchanges χ_i and χ_j and fixes all the other χ_k. An interesting exercise then shows that W must be all of S_n, since otherwise Z^W would be too large. The last step here involves identifying the roots. But these must be of the form χ_i/χ_j since we know the reflections. But the Weyl group acts transitively on the set $\{\chi_i/\chi_j\}$, and so $\Phi = \{\chi_i/\chi_j \mid i \neq j\}$. The end result here is that the triple $(X(T), \Phi, X(Z))$ is what you get from $M_n(K)$. It follows easily from this (using 5.2) that $M \cong M_n(K)$.

Remark 5.6. A reductive monoid M is called **semisimple** if it is normal, has a one dimensional center and a zero element. Such a monoid is classified in [104] by its **polyhedral root system**

$$M \rightsquigarrow (X(T), \Phi, X(Z))$$

where T is a maximal torus of G with roots $\Phi \subseteq X(T)$, and Zariski closure $Z \subseteq M$. It follows from Theorem 5.4, that *all* reductive, normal monoids are classified by such triples.

Remark 5.7. One can characterize all smooth reductive algebraic monoids in the spirit of Theorem 5.5. We refer the interested reader to the work of Timashev [138].

5.2 Vinberg's Approach

In this section we describe another approach, due to Vinberg [142], to the classification of reductive monoids. In this approach, we assume that K is an algebraically closed field of characteristic zero. This has some obvious advantages which we exploit (following [142]). On the other hand, Rittatore [123] has since proved that this assumption on K is not really necessary for many of Vinberg's results. For simplicity, we stick to Vinberg's approach.

Let M be a reductive, normal monoid with unit group G. We obtain

$$K[M] \subseteq K[G]\,.$$

Now, it is well known that

$$K[G] = \bigoplus_{\lambda \in X_+} K[G]_\lambda$$

where X_+ is the set of dominant characters of T with respect to B. Here, each $K[G]_\lambda$ is an irreducible $G \times G$-module with highest weight $\lambda \otimes \overline{\lambda}$ and $G \times G$ acts on G via $((g,h),x) \mapsto gxh^{-1}$. This "multiplicity ≤ 1" condition implies that any $G \times G$-stable subspace of $K[G]$ is a sum of some of the $K[G]_\lambda$. In particular,

$$K[M] = \bigoplus_{\lambda \in L(M)} K[G]_\lambda$$

where $L(M) \subseteq X_+$. We refer to $L(M)$ as the **augmented cone** of M. It is clear that distinguishing M from the other reductive monoids with unit group G amounts to identifying $L(M)$. Notice also that $L(M)$ is not a cone in the usual sense, rather it is the set of lattice points of such a cone.

Definition 5.8. *a) For $\lambda, \mu \in X_+$ define $X(\lambda, \mu)$ as the set of highest weights of the irreducible components of $K[G]_\lambda K[G]_\mu$. Thus,*

$$K[G]_\lambda K[G]_\mu = \bigoplus_{\nu \in X(\lambda,\mu)} K[G]_\nu \ .$$

It is known that $\lambda + \mu \in X(\lambda, \mu)$ and also that, if $\nu \in X(\lambda, \mu)$, then

$$\nu = \lambda + \mu - \sum k_i \alpha_i$$

where $k_i \geq 0$.
b) An additive submonoid $L \subseteq X_+$ is called perfect *if*

$$\lambda, \mu \in L \quad \textit{implies} \quad X(\lambda, \mu) \subseteq L \ .$$

Theorem 5.9. *A submonoid $L \subseteq X_+$ defines an algebraic monoid with unit group G if and only if it is perfect, finitely generated, and it generates $X(T)$ as a group.*

Proof. By definition, any subspace of $K[G]$ of the form $K[G]_L = \bigoplus_{\lambda \in L} K[G]_\lambda$, for $L \subseteq X_+$ perfect, is a subalgebra of $K[G]$. On the other hand, one checks that $\triangledown(K[G]_\lambda) \subseteq K[G]_\lambda \otimes K[G]_\lambda$ for any $\lambda \in X_+$. Hence $K[G]_L$ defines a monoid M. Now $K[G]_L$ is finitely generated by results of [76] and so M is algebraic. To show that $K[G]_L = K[M]$ and $K[G]$ have the same quotient field one needs to know that the representations $\rho_\lambda : G \longrightarrow G\ell(V_\lambda)$, $\lambda \in L$, separates the points of G (and conversely). See [142] for the details.

Vinberg goes on to characterize those monoids, as above, with unit group G which are normal. This makes use of a result, due to Popov in [76], that implies $K[M]$ is normal if and only if $K[M]^{U^- \times U}$ is normal. This allows a decisive characterization in terms of rational polyhedral cones related to $X(T) \otimes \mathbb{Q}$ and its dual. We describe here Vinberg's classification of normal reductive monoids.

Theorem 5.10. *A subset $L \subseteq X_+$ defines a normal algebraic monoid $M(L)$ with G as unit group if and only if $L = X_+ \cap K$, where $K \subseteq X(T) \otimes \mathbb{Q}$ is a closed convex polyhedral cone such that*

a) $-\Delta = \{-\alpha_1, \dots, -\alpha_n\} \subseteq K$.
b) $K \cap \mathcal{C}$ *generates* $X(T) \otimes \mathbb{Q}$, *where* $\mathcal{C}$ *is the Weyl chamber of* T.

The reader is refered to Theorem 2 of [142] for many details. Rather than reproducing those proofs here, we show how the invariant K can be obtained from the results of § 5.1. Conversely, we also indicate how Theorem 5.4 of § 5.1 can be deduced from Theorem 5.10.

Notice that any cone L as in Theorem 5.10, automatically defines a reductive monoid $M = M(L)$ (using Theorem 5.9) since, if $\nu \in X(\lambda, \mu)$, we obtain

$$\nu = \lambda + \mu - \sum k_i \alpha_i$$

where $k_i \geq 0$, for all i. Furthermore $M(L)$ has a zero element if and only

a) $K \cap X(T/T_0)$ is a pointed cone, and
b) $K \cap \mathcal{C}_0 = \{0\}$, where $\mathcal{C}_0$ is the Weyl chamber of G_0.

There is a unique largest such K with $\mathcal{C} \cap K = L$. Notice however, that Vinberg works over $\mathbb{Q}$.

Given the augmented cone $L = L(M)$, with M normal and reductive, one can view $L(M) \subseteq X(\overline{T})$ as a fundamental domain for the Weyl group action on $X(\overline{T})$. Since $X(\overline{T})$ is an integral polyhedral cone, we obtain

$$X(\overline{T}) = \cap_{\nu \in V^1} X(T)_\nu$$

where

a) $V^1 = \{\nu_e : X(T) \to \mathbb{Z} \mid \nu_e$ is induced from $\lambda_e : K^* \subseteq T\}$. Here $\lambda_e : K^* \subseteq T$ is the unique $1-PSG$ that converges to the idempotent $e \in E^1(\overline{T})$;
b) $X(T)_\nu = \{\chi \in X(T) \mid \nu(\chi) \geq 0\}$.

Now for each $f \in E^1(\overline{T})$, there is a unique $e \in \Lambda^1 = \Lambda \cap E^1(\overline{T})$ such that $e = wfw^{-1}$ for some $w \in W$. One observes that

$$e \rightsquigarrow \nu_e$$

identifies Λ^1 with

$$N^1 = \{\nu \in V^1 \mid \nu(-\alpha) \geq 0 \text{ for all } \alpha \in \Delta\}.$$

Hence the set of integral points of the maximal K (for this L) is

$$K(\mathbb{Z}) = \bigcap_{e \in \Lambda^1} X(T)_{\nu_e}.$$

We can now assess how the classification of normal monoids in [142] is related to the classification theory of § 5.1. Let $K(\mathbb{Q})$ denote Vinberg's K.

Corollary 5.11. *Let M be normal. Let $K(\mathbb{Q})$ and $X(\overline{T})$ be as above. Then*

a) $X(\overline{T}) = (\cap_{w\in W} w(K)) \cap X(T)$;
b) $K(\mathbb{Q}) = \cap_{e\in\Lambda^1}(X(T)\otimes\mathbb{Q})_{\nu_e}$.

5.3 Algebraic Monoids as Spherical Varieties

The theory of reductive algebraic monoids can be thought of as a significant special case within the theory of spherical embeddings [121]. Indeed, the action

$$\mu : G\times G\times G \longrightarrow G$$
$$\mu((g,h),x) = gxh^{-1}$$

proves that G is isomorphic to $(G\times G)/\Delta G$, which is homogeneous for $G\times G$. Furthermore, the Borel subgroup $B\times B^-\subseteq G\times G$ has a dense orbit on G. This is the key assumption that makes the theory of spherical varieties work. Thus any algebraic monoid M with unit group G is a spherical embedding for $(G\times G)/\Delta G$.

5.3.1 Spherical Varieties

Spherical embeddings are classified using a numerical set-up known as the **coloured cone**. This theory was founded by Luna and Vust in their seminal paper [56]. Rittatore [121] has identified how reductive monoids fit into this more general general set-up, as we shall see in the next section.

We summarize briefly this beautiful theory. The reader who wants to pursue this more general point of view in detail should consult [10, 11, 48, 56, 55, 139, 144].

Let G be reductive and let G/H be a **spherical homogenous space** for G. Then by definition, a Borel subgroup B of G has a dense orbit on G/H. Let $G/H\subseteq X$ be a **simple embedding** of G/H. Then by definition, G acts on X, X has a unique closed G-orbit $Y\subseteq X$, and X is normal.

Now let

$$\Omega = \left\{\chi\in X(B)\middle| \begin{array}{l} g\cdot f = \chi(g)f \text{ for all } g\in B \text{ and}\\ \text{some } f\in K(G/H) \text{ with } f\neq 0\end{array}\right\}.$$

The key point here is that, if f_1, f_2 are two nonzero rational functions on G/H of weight $\chi\in\Omega$, then $f_1/f_2\in K(G/H)^B = K$. Therefore any discrete valuation ν over K has the property $\nu(f_1)=\nu(f_2)$. Hence

$$\nu(\chi)\in\mathbb{Q} \text{ is well defined for } \chi\in\Omega\,.$$

In particular, we may think of G-invariant, discrete valuations on $K(G/H)$ as elements of

$$Q(G/H) = Hom(\Omega, \mathbb{Q}) .$$

Let ρ_ν be the element of $Q(G/H)$ associated with the G-invariant valuation ν of $K[G/H]$. It is a fundamental theorem of Brion and Vinberg [9, 141] that B has only a finite number of orbits on any spherical embedding. The original proof of this theorem applied only in characteristic zero as it relied heavily on the results of Popov in [76]. However, Grosshans [35] has since extended many of the results of [76] to arbitrary characteristic. Thus, the general proof of this fundamental finiteness result in positive characteristic is similar to Popov's original proof.

Definition 5.12. *Let G and G/H be as above, and let X be simple embedding of G/H with unique closed orbit Y.*

a) Denote by $\mathcal{V}(G/H)$ the set of G-invariant discrete valuations of $K[G/H]$.
b) Denote by $\mathcal{D}(G/H)$ the (finite) set of B-stable prime divisors of G/H.
c) Let $\mathcal{F}(X) = \{D \in \mathcal{D} \mid Y \subseteq \overline{D}\}$.
d) $\mathcal{B}(X) = \{D \subseteq X \mid D$ is an irreducible G-stable divisor of $X\} \subseteq Q(G/H)$. The inclusion is obtained by first identifying $D \in \mathcal{B}$ with its G-invariant, discrete, normalized valuation on $K[G/H]$, and then applying the above remarks.

$\mathcal{F}(X)$ is called the set of colors *of X.*

Remark 5.13. a) The map $\nu \to \rho_\nu$ is injective.
b) Each $D \in \mathcal{D}(G/H)$ determines a valuation ρ_D of $K[G/H]$. The map $D \to \rho_D$ may be noninjective in general. However, it is injective for reductive monoids. See part b) of Proposition 5.15 below.

The basic theorem here ([48]) is as follows.

Theorem 5.14. *Let $G/H \subseteq X$ be a simple embedding with closed orbit $Y \subseteq X$. Let $\mathcal{F}(X)$ and $\mathcal{B}(X)$ be as above. Let $\mathcal{C}(X)$ be the rational cone in $Q(G/H)$ generated by $\mathcal{B}(X) \cup \rho(\mathcal{F}(X))$. Then the correspondence*

$$X \longmapsto (\mathcal{C}(X), \mathcal{F}(X))$$

uniquely determines the simple normal G-embedding X of G/H to within G-isomorphism.

The complete and correct formulation of Theorem 5.14 requires the definition of a **colored cone** [48]. That way one can characterize axiomatically exactly which pairs $(\mathcal{C}, \mathcal{F})$ can arise from some spherical embedding X of G/H. However, it is not our mission here to write the book on spherical embeddings. Theorem 5.14 implies, in particular, that X is determinied to within isomorphism by

$$(\mathcal{B}(X), \mathcal{F}(X)).$$

This is sufficient for our purposes since, in the case of reductive normal monoids, the exact classification has been made precise in Theorem 5.4. See also Theorem 5.16 below.

5.3.2 Rittatore's Approach

Let G be a reductive group. We regard G as the homogeneous space for $G \times G$ defined by the action

$$\mu : G \times G \times G \to G$$

$\mu((g,h),x) = gxh^{-1}$. As already mentioned, the Bruhat decomposition shows us that G is spherical for $G \times G$. Indeed, the Borel subgroup $B \times B^-$ of $G \times G$ has a finite number of orbits on G. Hence exactly one of them must be dense in G. For convenience, we simply write G for $(G \times G)/\Delta(G)$, where $\Delta(G) = \{(g,h) \in G \mid g = h\}$. Let $B \subset G$ be a Borel subgroup of G, and let B^- be the Borel subgroup of G opposite to B relative to the maximal torus T of G. For convenience, we use the Borel subgroup $B \times B^-$ of $G \times G$ as the prefered Borel subgroup.

In order to describe Rittatore's work on reductive monoids, it will be necessary to identify the salient ingredients. These are:

a) the colors of G,
b) $\mathcal{Q}(G)$, and
c) $\mathcal{V}(G)$.

Let $X(T)$ be the set of characters of T, and let S be the set of simple reflections (of the Weyl group W) relative to T and B. Denote by s_α the simple reflection associated with the simple root α. Let $C(G)$ be the Weyl chamber of G associated with T and B.

Proposition 5.15. *a) $\mathcal{Q}(G) \cong X(T)^* \otimes \mathbb{Q}$, the dual of $X(T) \otimes \mathbb{Q}$.*
b) The colors of G are the $B \times B^-$-invariant divisors $\{D_\alpha = \overline{Bs_\alpha B^-}\}$ as α ranges over the simple roots. The valuation ρ_D (associated with $D = D_\alpha \in \mathcal{D}$) is determined in $\mathcal{Q} = X(T)^* \otimes \mathbb{Q}$, by the rule $\rho_D = \alpha^\vee \in X(T_0)^* \otimes \mathbb{Q}$.
c) $\mathcal{V}(G) = -C(G) \subseteq \mathcal{Q}(G)$

Proof. The set of weights Ω of $K(G)^{(B \times B^-)}$ is canonically isomorphic to $X(T)$. Hence

$$Hom_{\mathbb{Z}}(\Omega, \mathbb{Z}) \cong X(T)^*.$$

It is well known that the codimension one orbits of the Bruhat decomposition are as stated. See Proposition 9 of [121] for the calculation of ρ_D.

The proof of c) is more complicated, and consequently we refer the reader to [121] (especially Proposition 8 of [121]) for most of the details. However, the basic idea is easy to describe, and we do that here. Associated with each normalized $G \times G$-valuation $\nu \in \mathcal{V}(G)$ is an *elementary embedding* U of G (as a $G \times G$-variety). Choose a $1 - PSG$ $\lambda \in X(T)^*$ such that $lim_{t \to 0}\lambda(t) = x$ exists and belongs to $U \backslash G$. We can assume that $\lambda \in -C(G)$, by conjugating λ if necessary. It then follows that BxB^- is open in $U \backslash G$. It follows from this that ν is equivalent to the valuation ν_λ defined as follows.

K^* acts on G via λ through $G \times G$. Denote the induced action on $K[G]$ by $t \cdot f = \lambda(t)^*(f)$. This determines a decomposition $K[G] = \oplus_{n \in \mathbb{Z}} K[G]_n$, where

$$K[G]_n = \{f \in K[G] \mid t \cdot f = t^n f\}.$$

For $f \in K[G]$, we can write $f = \sum f_n$, so we define $\nu_\lambda(f) = inf\{n \mid f_n \neq 0\}$. It is easy to check that ν_λ extends to a valuation of $K(G)$. One needs to check that ν is $G \times G$-invariant. For those details we refer the reader to [121].

We now let M be a normal monoid with reductive unit group G. Assume for simplicity that M has a zero element (The general case is only superficially more complex.). Then

$$\mathcal{F}(M) = \mathcal{D}(G)$$

the set of all $B \times B^-$-stable, irreducible divisors of G. Thus, all colors of G are involved in M, and so not surprisingly they do not play much of a rôle in the classification of reductive monoids with zero. Thus, we see that M is determined by $\mathcal{C}(M)$, or what amounts to the same thing, $\mathcal{B}(M)$.

We can now paraphrase Rittatore's identification (Theorem 4 of [121]) of the theory of reductive monoids within the theory of spherical embeddings.

Theorem 5.16. *Let G be a reductive group. The irreducible, normal algebraic monoids M with unit group G are in bijective correspondence with the strictly convex polyhedral cones in $\mathcal{Q}(G)$ generated by all of $\mathcal{D}(G)$ and a finite set of elements of $\mathcal{V}(G)$.*

The following proposition (essentially a special case of Proposition 13 of [121]) indicates how Rittatore's cone is the dual of Vinberg's cone.

If the solvable group B acts rationally on the K-algebra A, we denote by $A^{(B)}$, the subset of B-eigenvectors of the action.

Proposition 5.17. *Let M be normal with group G, and zero element $0 \in M$. Then*

$$K[M]^{(B \times B^-)} = \{f \in K(G)^{(B \times B^-)} \mid \chi_f \in \mathcal{C}(M)^\vee\}.$$

Thus if $char(K) = 0$ then

$$K[M] = \bigoplus_{\lambda \in \mathcal{C}(M)^\vee} K[G]_\lambda.$$

Hence

$$\mathcal{C}(M)^\vee = L(M)$$

where $L(M)$ is the augmented cone used in § 5.2 by Vinberg.

Proof. The first equality follows from Theorem 3.5 of [48]. Everthing else here can be deduced from this. On the other hand, by part b) of Corollary 5.11, $K = \cap_{e \in \Lambda^1}(X(T) \otimes \mathbb{Q})_{\nu_e}$, while $X(T)_+ = \{\chi \in X(T) \mid \nu_\alpha(\chi) \geq 0 \text{ for all } \alpha \in \Delta\}$, and so $L(M) = K(\mathbb{Z}) \cap X(T)_+$. But $K(\mathbb{Z}) \cap X(T)_+ = \mathcal{C}^\vee$.

The results of Proposition 5.17 hold in arbitrary characteristic, since any normal monoid has a good $G \times G$-filtration. See Definition 9.7.

5.3.3 Type Maps and Colors

Let $Y = GyG$ be a $G \times G$-orbit of the reductive monoid M. Let

$$X = \{x \in M \mid Y \subseteq \overline{GxG}\}.$$

A description of $(\mathcal{B}(X), \mathcal{F}(X))$ can be obtained directly from the theory of spherical embeddings. However, there is also a monoid approach.

In this subsection, we take the following point of view in the study of the orbit structure of a reductive monoid. On the one hand, as in the previous subsection, we obtain,

a) $\mathcal{D}(G) = \mathcal{F}(M)$, the colors of M, and
b) $\mathcal{B}(M)$, the set of $G \times G$-invariant valuations of G associated with M (identified with the appropriate 1-PSGs of G).

On the other hand, we obtain the type map, (see Definition 4.6).

$$\lambda : \Lambda \to 2^S$$

which is the combinatorial invariant of M analagous to the Dynkin diagram. $\Lambda = \{e \in E(\overline{T}) \mid eB = eBe\}$, the cross-section lattice (relative to B^-), is a set of idempotent representatives for the set of $G \times G$-orbits of M. The type map says a lot about how the $G \times G$-orbits of M fit together to make the monoid structure of M possible.

In this subsection, we continue our comparison of the two viewpoints, and further assess how the type map approach can be described in terms of the "divisors, colors, and cones" approach.

Let M be a reductive monoid with 0, and recall that

$$\Lambda^1 = \{e \in \Lambda \backslash \{1\} \mid e \text{ is maximal}\}.$$

Λ^1 is the algebraic monoid notion equivalent to $\mathcal{B}$. Indeed, for each $e \in \Lambda^1 \subseteq E(\overline{T})$, there is an essentially unique 1-PSG $\nu : K^* \to T$ such that $lim_{t \to 0}\nu(t) = e$. So we can make the identification of

$$\mathcal{B}(M) \cong \Lambda^1$$

where it is understood that Λ^1 includes the set of 1-PSGs involved.

Notice that, for $e \in \Lambda$,

$$\lambda(e) = \lambda^*(e) \sqcup \lambda_*(e),$$

where $\lambda^*(e) = \{s \in S \mid se = es \neq e\}$ and $\lambda_*(e) = \{s \in S \mid se = es = e\}$. See Definition 7.11.

Theorem 5.18. *Let $e \in \Lambda$ and let $s \in S$. Then $s \in \lambda_*(e)$ if and only if $GeG \subseteq \overline{BsB^-}$.*

Proof. Now BeB^- is dense in GeG (since $BB^- \subseteq G$ is dense), and so $GeG \subseteq \overline{BsB^-}$ if and only if $BeB^- \subseteq \overline{BsB^-}$, and this happens if and only if $e \in \overline{BsB^-}$. Equivalently, $e \in e\overline{BsB^-}e$. But $e\overline{BsB^-}e = \overline{eBeseB^-e}$. But H_e is a reductive group with opposite Borel subgroups eBe and eB^-e, and Weyl group $W^*(e) = eC_W(e)e$. Furthermore, $eBeB^-e$ is the dense $eBe \times eB^-e$-orbit of H_e. Thus, $e \in \overline{eBeseB^-e}$ is equivalent to saying that $e = ese$. This is equivalent to $es = se = e$ for, if e and s do not commute, ese must be strictly less than e in the $\mathcal{J}$-order.

For an algebraic monoid M we make the following definition.

Definition 5.19. *For $e \in \Lambda$ define*

$$\lambda^1(e) = \{f \in \Lambda^1 \mid fe = ef = e\}.$$

The following result helps to illustrate how the Luna-Vust approach to spherical varieties applies to the case of reductive monoids. We give an algebraic monoid style proof.

Proposition 5.20. *Let M be reductive, and let $e \in \Lambda$. Then e is uniquely determined by*

$$(\lambda^1(e), \lambda_*(e)) \in 2^{\Lambda^1} \times 2^S.$$

Furthermore, $e \leq f$ if and only if $\lambda_*(e) \subseteq \lambda_*(f)$ and $\lambda^1(e) \subseteq \lambda^1(f)$.

Proof. Consider e as an element of $E(\overline{T})$, and let $A = \{f \in E^1(\overline{T}) \mid fe = ef = e\}$, where $E^1(\overline{T})$ is the set of maximal idempotents of $E(\overline{T})\backslash\{1\}$. Then

$$e = \Pi_{f \in A} f.$$

But it is easy to see that

$$\{f \in E^1(\overline{T}) \mid fe = ef = e\} = \{wfw^{-1} \mid f \in \lambda^1(e) \text{ and } w \in W_*(e)\}.$$

Thus e is determined by $(\lambda^1(e), \lambda_*(e))$ since $W_*(e)$ is the Coxeter group generated by $\lambda_*(e)$. The claim is clear from this.

For the reductive monoid M, consider the $G \times G$-orbit $Y = GeG$, where $e \in \Lambda$, and define

$$X = \{x \in M \mid Y \subseteq \overline{GxG}\}.$$

X is the unique open $G \times G$-subvariety of M with Y as its only closed orbit. As in Theorem 5.14, X is determined up to isomorphism by $(\mathcal{B}(X), \mathcal{F}(X))$. Proposition 5.20 above allows the following identification:

$$(\mathcal{B}(X), \mathcal{F}(X)) \cong (\lambda^1(e), \lambda_*(e))$$

where, as above, it is understood that each element of $\lambda^1(e)$ "remembers" its associated $G \times G$-invariant valuation of $K[G]$.

It is desirable to have an identification of the colors of the spherical $G \times G$-homogeneous spaces GeG, $e \in \Lambda$. Indeed, in many cases, the boundary of $\overline{GeG}$ has codimension greater than one in $\overline{GeG}$. Hence in those cases, $\overline{GeG}$ is determined as a $G \times G$-variety, by its colors.

Theorem 5.21. *Let M be reductive, and let $e \in \Lambda$. Then*

a) $GeG = \sqcup_{r \in W_eW} BrB$.
b) The codimension one $B \times B^-$-orbits of GeG are as follows. There are three types.
  i) $BesB^- = BseB^-$, *where $s \in S$ and $es = se \neq e$.*
  ii) $BesB^-$, *where $s \in S$ and $es \neq se$.*
  iii) $BseB^-$, *where $s \in S$ and $se \neq es$.*
  In each case, the $B \times B^-$-orbit in question is contained in $\overline{BsB^-}$.

Proof. a) follows from Theorem 8.8. The list of codimension one $B \times B^-$-orbits in b) follows from Theorem 14.1, taking into account that $B^- = w_0Bw_0^{-1}$, where w_0 is the longest element in the Weyl group.

$\overline{BsB^-}$ is stable under the action of T (on the left or the right). Thus, by continuity, $e\overline{BsB^-} \subseteq \overline{BsB^-}$. Similarily, $\overline{BsB^-}e \subseteq \overline{BsB^-}$. The last claim in b) follows easily from this.

Example 5.22. Let $M = M_n(K)$. Then M is reductive with unit group $G = G\ell_n(K)$. Let $B \subseteq G$ be the Borel subgroup of invertible lower-triangular matrices. Let B^- be the invertible upper-triangular matrices. For $e \in E(M)$ define (as in [83])

$$\varphi_e(x) = \det(exe + 1 - e)$$

for $x \in M$. Let

$$s_i = \begin{pmatrix} 1 & & & & & & \\ & \ddots & & & & & \\ & & 1 & & & & \\ & & & 0 & 1 & & \\ & & & 1 & 0 & & \\ & & & & & 1 & \\ & & & & & & \ddots \\ & & & & & & & 1 \end{pmatrix} \qquad i = 1, \ldots, n-1$$

and

$$e_i = \begin{pmatrix} 1 & & & & & \\ & \ddots & & & & \\ & & 1 & & & \\ & & & 0 & & \\ & & & & \ddots & \\ & & & & & 0 \end{pmatrix} \qquad i = 1, \ldots, n-1,$$

so that rank $(e_i) = i$. Then $\mathcal{D}(G) = \{Bs_iB^- \mid i = 1, ..., n-1\}$ and $\overline{Bs_iB^-} = \varphi_{e_i}^{-1}(0)$.

Let $X_i = \{x \in M \mid \text{rank}(x) \geq i\}$. Then

$$\mathcal{F}(X_i) = \lambda_*(e_i) = \{s_{i+1}, \ldots, s_{n-1}\}.$$

6

Universal Constructions

Every branch of algebra has universal constructions of interest. If reductive monoids are involved, there are some pleasant surprises.

6.1 Quotients

Theorem 6.1. *Let M be an irreducible, algebraic monoid with unit group G, and let $R_u(G) \subseteq G$ be the unipotent radical of G. Then there exists a unique algebraic monoid N, and a surjective morphism $\pi : M \to N$ such that*

a) the unit group of N is $G/R_u(G)$;
b) $\pi|G$ is the usual quotient morphism from G to $G/R_u(G)$;
c) π is universal for morphisms from M with kernel containing $R_u(G)$.

Furthermore, if $Z \subseteq M$ is a maximal D-submonoid of G, then $Z' = \pi(Z)$ is a maximal D-submonoid of N, and $\pi : Z \to Z'$ is an isomorphism.

Proof. By Theorem 3.8 we can assume that $M \subseteq End(V)$ is a closed submonoid. There exists a filtration

$$0 = V_1 \subseteq V_2 \subseteq \cdots \subseteq V_n$$

of M-modules such that V_i/V_{i-1} is a simple M-module. Hence there is a canonical (induced) morphism

$$\varphi : M \to \Pi_{i=2}^{n} End(V_i/V_{i-1}).$$

Let $N_1 = \overline{\varphi(M)}$. Then N_1 is an irreducible algebraic monoid with unit group $G(N)$ abstractly isomorphic to $G/R_u(G)$. Furthermore, φ induces an isomorphism $\varphi : Z \to Z'$ on maximal D-submonoids, since we are just passing to the associated graded object. By part c) of Theorem 4.5, $N_1 = G(N)Z'G(N)$, and so $\varphi : M \to N_1$ is surjective. However, N_1 may not be quite right. So define

$$N = M/U = \text{the normalization of } N_1 \text{ along } G/R_u(G) \to G(N_1).$$

Corollary 6.2. *Let M be an irreducible, algebraic monoid with solvable unit group $G = TU$. Then the coordinate ring $\mathcal{O}[M/U] \subseteq \mathcal{O}[M]$ of M/U is $K[X(M)]$, the semigroup algebra of the set of characters of M.*

Proof. In this case, $N = M/U$ is a D-monoid. Hence $\pi : Z \to N$ is an isomorphism by Theorem 6.1, and $\mathcal{O}(M/U) = K[X(M)]$, the monoid algebra, since M/U is a D-monoid.

In group theory, the **abelization** is already available abstractly. Therefore the corresponding results for algebraic groups are not so surprising. For monoids, the situation is somewhat different. The following result is mostly due to Vinberg [142].

Theorem 6.3. *Let M be a normal reductive algebraic monoid with unit group G. Let G_0 be the semisimple part of the unit group of G. Let A be the geometric invariant theory quotient of M by the action $(g, h, x) \to gxh^{-1}$ of $G_0 \times G_0$ on M. Let $\pi : M \to A$ be the quotient morphism. Then*

a) the coordinate algebra of A is $K[X(M)]$, the monoid algebra of the character monoid of M;
b) if Z is the connected center of G, then $\pi : \overline{Z} \to A$ is the finite dominant quotient morphism obtained from the action of $Z \cap Z(G_0)$ on $\overline{Z}$;
c) the $G_0 \times G_0$-orbit $G_0xG_0 \subseteq M$ is closed if and only if it intersects $\overline{Z}$.

Proof. Assume that M has a zero element. The general case is only superficially more complicated. Now the semisimple part G_0 of any reductive monoid is closed in M (proof: $G_0 = \cap_{\chi \in X(M)} ker(\chi)$). Thus, for any $e \in E(\overline{Z})$ $eG_0 \subseteq eM$ is closed, since eG_0 is the semisimple part of the reductive monoid eM. But $eG_0 = G_0eG_0 = G_0e$. Hence by the basic theorem of geometric invariant theory,

$$\pi|\overline{Z} : \overline{Z} \to A$$

separates the idempotents of $\overline{Z}$. Let $Z_0 = Z \cap Z(G_0)$. It follows that $\pi(\overline{Z}) \subseteq A$ is open and $G(A)$-invariant. Since M has a zero element, so too does A, and $\pi(0_M) = 0_A$. Thus, $\pi(\overline{Z}) = A$. In any case,

$$\overline{Z}/Z_0 \to A$$

is finite, dominant and birational. Hence by Zariski's main thereom (Theorem 2.29), $\overline{Z}/Z_0 \to A$ is an isomorphism. But then $\overline{Z}$ must meet every closed $G_0 \times G_0$-orbit.

6.2 Class Groups of Reductive Monoids

Definition 6.4. *Let M be reductive. We say that M is* locally simply connected *(lsc) if $H(e)$ has trivial divisor class group for each $e \in \Lambda$.*

We are here using the word "locally" in the sense of semigroup theory, rather than in the sense of geometry. The property $\mathcal{P}$ of a semigroup S is called a **local property** if, for any idempotent $e \in S$, $\mathcal{P}$ holds for the semigroup eSe. Notice also that, strictly speaking, any simply connected group is semisimple.

Remark 6.5. The following results are discussed in detail in § 2.2.5. We state them here for convenience. Let G be a connected reductive group with commutator subgroup G'.

a) A connected, semisimple group G is simply connected if and only if $Cl(G) = 0$.
b) Suppose that G is a connected, reductive group whose commutator subgroup $G' = (G, G)$ has trivial divisor class group. Then the same property holds for any Levi subgroup L of G. (This also follows from Remark 2.13 and Lemma 2.17 of [137].)
c) For any connected reductive group G, there exists a connected reductive group G_1 with $Cl(G_1) = 0$, and a finite dominant morphism $\zeta : G_1 \to G$.
d) If G is connected and reductive, then $Cl(G) = 0$ if and only if $Cl(G') = 0$.

Definition 6.6. *M is* $\mathcal{J}$-coirreducible *if Λ^1 is a singleton.*

Recall that $\Lambda^1 \subseteq \Lambda$ is the subset of Λ that represents the codimension one $G \times G$-orbits of M. See part a) of Definition 4.6 and §5.3.3.

Theorem 6.7. *Let M be reductive.*

(a) If $Cl(M) = (0)$ then M is locally simply connected.
(b) If M is $\mathcal{J}$-coirreducible then there exits $M' \to M$, finite and dominant, such that $Cl(M') = (0)$. Furthermore, $Cl(M)$ is finite.
(c) For any such M there exists $\pi : M' \to M$ such that
  (i) $Cl(M') = (0)$ and
  (ii) π induces a bijection of $\mathcal{U}^1(M') \to \mathcal{U}^1(M)$, where $\mathcal{U}^1(M)$ denotes the set of maximal $\mathcal{J}$-classes of $M \backslash G$.

Proof. Notice first that, given X such that $Cl(X) = (0)$, then $Cl(U) = (0)$ for any open set $U \subseteq X$. Thus, to prove (a) it suffices to show that $Cl(eMe) = (0)$ for any $e \in E(M)$. We apply Theorem 10.6 of [29]. This theorem says that if the K-algebra $A = \oplus_{i \geq 0} A_i$ is a graded, noetherian, and factorial then A_0 is also factorial. To do this we must show that $A = K[M] = \oplus_{n \geq 0} A_n$ with $A_0 = K[eMe]$. But we know from Corollary 6.10(ii) of [82] that there exists a one parameter subgroup $\lambda : K^* \to G$ such that $\lim_{t \to 0} \lambda(t) = e$. Consider the action $\mu : K^* \times M \to M$ given by $\mu(t, x) = \lambda(t)x\lambda(t)$. μ induces a rational action $\rho : K^* \to Aut(K[M])$. If we let $A_n = \{f \in K[M] | \rho(t)(f) = t^n f\}$ then $K[M] = \oplus_{n \geq 0} A_n$ is the desired $\oplus$-decomposition.

To prove (b) we may assume without loss of generality that $0 \in M$. Notice that this implies that $dim\ Z(G) = 1$. Now $M \backslash G$ is irreducible of codimension one in M. So it determines a divisor class $D \in Cl(M)$. Since $M \backslash G$ is irreducible, we have by part c) of Proposition 2.19 an exact sequence

$$0 \to \mathbb{Z} \cdot D \to Cl(M) \to Cl(G) \to 0. \tag{1}$$

By Theorem 3.4 of [101], there exists $\chi : M \to k$ such that $\chi^{-1}(0) = M \backslash G$. Hence, in $Cl(M)$, D has finite order. We conclude that $Cl(M)$ is finite since, from Corollary 2.8 of [42], $Cl(G)$ is finite. To find M' we first assume that $Cl(G) = (0)$. For if $Cl(G) \neq 0$, we first consider $\zeta : G_1 \to G$, a finite dominant morphism with $Cl(G_1) = (0)$. We then apply Lemma 7.1.1 of [104] to obtain a reductive monoid M_1 with unit group G_1 and a finite dominant morphism $M_1 \to M$ extending ζ. With another application of Lemma 7.1.1 of [104] we may assume that $G = G_0 \times K^*$ where G_0 is semisimple and simply connected. From the exact sequence in (1) we obtain $Cl(M) = \mathbb{Z} \cdot D$, a finite cyclic group. Before we construct M' we need to determine exactly what controls the order of D in $Cl(M)$. Let $e \in \Lambda \backslash \{1\}$ be the unique maximal element, and let $T_e = T \cup eT = \overline{T}$. Then $T_e \subseteq \overline{T}$ is an open submonoid. Furthermore, by Lemma 5.1 there exist opposite Borel subgroups B, B^- containing T such that $m : B_u^- \times T_e \times B_u \to M$, $m(x, y, z) = xyz$, is an open embedding. Letting $R = K[M]$ and $S = K[B_u^- \times T_e \times B_u]$, we obtain $R \subseteq S$. Since $T_e \cong (K^*)^{r-1} \times K$ as varieties, we see from Corollary 7.2 and Theorem 8.1 of [29] that S is a UFD. Let $\mu = \{f \in R | f|M \backslash G = 0\}$. Clearly $|Cl(M)| = inf\{n | \mu^{(n)})$ is a principal ideal$\}$ where $\mu^{(n)}$ denotes the nth symbolic power of the ideal μ ([29]). So write $\mu^{(n)} = (\chi)$ where $n = |Cl(M)|$. We may assume that $\chi : M \to K$ is a morphism of algebraic monoids, adjusting the initial χ with a non-zero scalar if necessary. Consider $\chi \circ m \in S$. From our remarks above we see that $(\chi \circ m) = \mathfrak{p}^n$ where $\mathfrak{p} = \mu \cdot S$. Using the isomorphism $T_e \cong (K^*)^{r-1} \times K$ and the fact that $\chi \circ m$ factors through $p_2 : B_u^- \times T_e \times B_n \to T_e$, $(x, y, z) \rightarrowtail y$, we obtain the following diagram:

$$K \underset{j}{\hookrightarrow} T_e \underset{i}{\hookrightarrow} B_u^- \times T_e \times B_u \underset{\chi \circ m}{\longrightarrow} K.$$

j is the unique inclusion with the property $j(0) = e$. It follows that $n =$ degree $(\chi \circ m \circ i \circ j)$. Now let (X, ϕ, C) be the *polyhedral root system* of M (Definition 3.6 of [104]) and let $v : C \to N$ be the "valuation" determined by j (notice that $v^{-1}(0) \subseteq C$ is the facet of C determined by e). Let $\chi \in C$ denote the restriction of χ to $\overline{T}$. We can construct a new polyhedral root system (X', ϕ', C') as follows:

Since $G = G_0 \times K^*, X = X_0 \oplus \mathbb{Z}$. Furthermore, $C \subseteq X_0 \oplus \mathbb{N}$ and $\chi = (0, 1) \in C$. We define

$$X' = X_0 \oplus \frac{1}{n}\mathbb{Z}$$
$$\phi' = \phi$$
$$C' = \{\zeta \in X' | m\zeta \in C \text{ for some } m > 0\}.$$

It is easily checked that (X', ϕ', C') is the polyhedral root system of the reductive monoid M' obtained via Lemma 7.1.1 of [104] from the map $\zeta : G \to G \subseteq M$ given by $\zeta(g, \alpha) = (g, \alpha^n)$. Furthermore, $v : C \to \mathbb{N}$ extends uniquely to $v' : C' \to \mathbb{N}$ via $v'(a, b/n) = v(a, 0) + b = v(a, 0) + (1/n)v(0, b)$. Notice that if $\chi' = (0,\ 1/n)$ then $v'(\chi') = 1$. But from our above calculation applied to M', $|Cl(M')| = v'(\chi')$. Hence $Cl(M') = (0)$.

For (c) we may assume M has a zero element. The general case is not essentially different. Let $e \in \Lambda^1$ be a maximal idempotent of $\Lambda\backslash\{1\}$. As above, there exists $v : C \to \mathbb{N}$ which extends to $v : X \to \mathbb{Z}$. Let

$$H = \{\chi \in X | v(\chi) \geq 0\}$$

and let

$$C_e = \cap_{w \in W} w^*(H) \subseteq X.$$

It is easily checked that (X, ϕ, C_e) is a polyhedral root system with $j : (X, \phi, C) \hookrightarrow (X, \phi, C_e)$. Let M_α be the associated reductive monoid, where $\alpha = GeG \in \mathfrak{U}^1(M)$. By construction M_α is $\mathfrak{J}$-coirreducible. Now from Theorem 8.1(a) of [104] there exists a birational morphism $\zeta_\alpha : M_\alpha \to M$ inducing j above. Applying part (b) above we can modify M_α slightly, if necessary, and assume that $Cl(M_\alpha) = (0)$. The unique maximal $\mathfrak{J}$-class of M_α gets mapped to α. After ordering $\mathfrak{U}^1(M)$, define

$$\zeta : \prod_{\alpha \in \mathfrak{U}^1(M_\alpha)} M_\alpha \to M$$

by $\zeta(x_1, \ldots, x_m) = \zeta_{\alpha_1}(x_1) \cdot \cdots \cdot \zeta_{\alpha_m}(x_m)$. Consider the action of $H = G_0 \times \underset{(m-1)}{\cdots} \times G_0$ on $N = \prod M_\alpha$ given by

$$(g_1, \ldots, g_{m-1})(x_1, \ldots, x_m) = (x_1 g_1^{-1}, g_1 x_2 g_2^{-1}, \ldots, g_{m-2} x_{m-1} g_{m-1}^{-1}, g_{m-1} x_m).$$

Define $M' = N/H$, the geometric invariant theory quotient of N by H, and let $q : N \to M'$ be the canonical quotient morphism.

By standard results of geometric invariant theory (Theorem 1.10 of [62]) there exists a unique morphism $\pi : M' \to M$ such that $\pi \circ q = m$, where $m : N \to M$ is the multiplication morphism. Based on Proposition 3.3 of [109] we see that M' is a reductive algebraic monoid with unit group $G_0 \times K^* \times \underset{(m)}{\cdots} \times K^*$. Furthermore, by Proposition 2.58, $Cl(M) = (0)$ since $Cl(N) = (0)$ and $G_0 \times \cdots \times G_0$ is a semisimple group. To complete the proof we must show that π identifies the maximal $\mathfrak{J}$-classes of M' with those of M. But π is surjective by construction, and so for each $\alpha \in \mathcal{U}^1(M)$ there exists $J \in \mathcal{U}^1(M')$ such that $\pi(J) = J_\alpha$. But also, $Cl(M') = (0)$, and so $|\mathcal{U}^1(M')| \leq dim(Z(G')) = m$. Thus $\mathcal{U}^1(M') \to \mathcal{U}^1(M)$ is bijective.

Remark 6.8. The above theorem was first recorded in [114]. Part c) leads directly to the construction of a "total coordinate ring" or "Cox ring" for the reductive monoid M. This generalizes some of the main results of [17]. See [28] for a general approach to this problem.

6.3 Flat Monoids

Vinberg [142] discovered a universal construction involving reductive monoids that has some remarkable properties. Let K be an algebraically closed field of characteristic zero. Associated with each semisimple group G_0 is a certain reductive monoid $Env(G_0)$ which is the universal flat deformation of G_0.

Recall from § 5.2, the augmented cone $L(M)$ of a reductive monoid M. Let $\overline{Z} \subseteq M$ be the closure in M of the connected center Z of G. It is easy to check that there is a natural partial order on $L(M)$ defined as follows.

$$\lambda_1 \geq \lambda_2 \text{ if } \lambda_1 = \chi\lambda_2$$

for some $\chi \in X(\overline{Z})$. Notice that this makes sense even though $X(Z) \not\subseteq L(M)$. Indeed, $X(Z) \otimes \mathbb{Q}^+ = X(M) \otimes \mathbb{Q}^+ \subseteq L(M) \otimes \mathbb{Q}^+$. Notice also that there may be a decomposition in $L(M)$ of the form $\lambda_1 = \chi\lambda_2$, with $\chi \in X(\overline{Z})$ but $\chi \notin X(M)$. The point here is that $T \cong T_0 \times_{Z_0} Z$, where $Z_0 = Z \cap T_0$. Thus $X(T) = \{(\lambda, \chi) \in X(T_0) \times X(Z) \mid \lambda|Z_0 = \chi|Z_0\}$. We let

$$\mathcal{M} \subseteq L(M)$$

denote the set of minimal elements of $L(M)$ with respect to the above partial order.

Before stating the main result, we need one more notion. As before, we let $X(T_0)_+$ denote the monoid of dominant weights of T_0. If $\lambda \in X(T_0)_+$, we can write

$$\lambda = \sum_{\alpha \in \Delta} c_\alpha \lambda_\alpha,$$

where $\{\lambda_\alpha\}$ is the set of fundamental dominant weights of G_0. Define

$$c : X(T_0)_+ \to Cl(M)$$

by $c(\lambda) = \sum_{\alpha \in \Delta} c_\alpha [\overline{Bs_\alpha B^-}]$. As in § 2.2.5, we write $D_\alpha = \overline{Bs_\alpha B^-}$. The main ideas behind the theorem below are due to Vinberg [142]. Our approach is a little different, in that we emphasise the rôle of the D_α.

Example 6.9. Let $M = M_2(K)$ and let $A = \begin{pmatrix} a & b \\ c & d \end{pmatrix} \in M$. Then $L(M) = \{(\delta^k, f_\lambda^l) \mid k > 0, l > 0\}$, relative to U and U^-, the upper-unitriangular and lower-unitringular groups of $Gl_2(K)$. Here $\delta(A) = ad - bc$, and $f_\lambda(A) = d$. Furthermore, λ is the fundamental dominant weight for $G_0 = Sl_2(K)$. Now

$$m : Z \times G_0 \to M$$

induces $j : L(M) \subseteq L(Z \times G_0) = \mathbb{Z}\chi \oplus \mathbb{N}f_\lambda(0)$, with $j(\delta) = \chi^2$ and $j(f_\lambda) = (\chi, f_\lambda(0))$. Hence

$$L(M) = \{(\chi^s, f_\lambda(0)^t) \mid s - t \geq 0,\ s - t \text{ is even}\}.$$

Then

$$\mathcal{M} = \{(\chi^t, f_\lambda(0)^t) \mid t \geq 0\}.$$

Notice that any $f \in L(M)$ can be written uniquely as

$$f = \delta^k f_\lambda^l = (\chi^{2k}, 1)(\chi^l, f_\lambda(0)^l).$$

Theorem 6.10. *Let M be a reductive monoid with unit group G, and let G_0 be the semisimple part of G. Assume (for convenience) that M has a zero element. The following are equivalent:*

a) The abelization $\pi : M \to A$ is flat, with reduced and irreducible fibres.
b) The following two conditions hold:
  i) If $\chi_1\lambda_1 = \chi_2\lambda_2$ ($\lambda_i \in \mathcal{M}$, $\chi_i \in X(\overline{Z})$) then $\chi_1 = \chi_2$ and $\lambda_1 = \lambda_2$.
  ii) $\mathcal{M}$ is a subsemigroup of $L(M)$.
c) The canonical map $c : X(T_0)_+ \to Cl(M)$ is trivial.
d) For any irreducible representation $\rho : M \to End(V)$ there is a character $\chi : \overline{Z} \to K$ of $\overline{Z}$, and an irreducible representation $\sigma : M \to End(V)$, such that $\sigma(e) \neq 0$ for any $e \in \Lambda^1$ and $\rho = \chi \otimes \sigma$.
e) Any $f \in L(M)$ factors as $f = \chi g$ where $\chi \in X(\overline{Z})$ and $g \in L(M)$ with zero set $Z(g) \subseteq \cup_{\alpha\in\Delta}\overline{Bs_\alpha B^-}$.

Proof. a) implies b). We sketch the proof from [142]. According to Proposition 3 of [142], if $\pi : M \to A$ is flat, then b)i) above holds. Now consider the inclusion $i^{-1}(0) \subseteq M$. Since π is flat, the induced map $i^* : K[M] \to K[\pi^{-1}(0)]$ induces an inclusion $i^* : \mathcal{M} \to K[\pi^{-1}(0)]\backslash\{0\}$. But $K[\pi^{-1}(0)]$ is an integral domain. Then $\mathcal{M}$ is multiplicatively closed and thus b)ii) holds.

b) implies a). By Proposition 3 of [142], condition b)i) implies that π is flat. If also b)ii) holds, then $K[\pi^{-1}(0)]^{U\times U^-} \cong K[\mathcal{M}]$ (semigroup algebra) via $i^* : K[M] \to K[\pi^{-1}(0)]$. Then $K[\pi^{-1}(0)]^{U\times U^-}$ has no zero divisors. Thus, by the results of [76], $K[\pi^{-1}(0)]$ has no zero divisors. This proves that $\pi^{-1}(0)$ is reduced and irreducible. But $Y = \{a \in A \mid \pi^{-1}(a) \text{ is reduced and irreducible}\}$ is open and G-invariant, by well known properties of morphisms [38]. Thus $Y = A$.

b) implies e). If b) holds, then we have already observed above that $i^*(\mathcal{M}) \subseteq K[\pi^{-1}(0)]\backslash\{0\}$. Furthermore, $K^* \cdot i^*(\mathcal{M}) = \{f \in K[\pi^{-1}(0)] \mid f \neq 0 \text{ and } BfB^- = K^*f\}$. We then let $f \in L(M)$. By our assumption, we have a unique factorization

$$f = \chi g$$

where $\chi \in X(\overline{Z})$ and $g \in (M)$. We must show that $Z(g) \subseteq \cup_{\alpha\in\Delta} D_\alpha$, where $D_\alpha = \overline{Bs_\alpha B^-}$. Recall from the proof of Theorem 5.2 that

$$M\backslash G = \cup_{e\in\Lambda^1} D_e,$$

where $D_e = \overline{GeG}$ and Λ^1 is the set of maximal idempotents of $\Lambda\backslash\{1\}$. Furthermore, if $\chi \in X(M)\backslash\{1\}$, then $Z(\chi) \subseteq \cup_{e\in\Lambda^1} D_e$. But $\pi^{-1}(0) = \cap_{\chi\in X(M)\backslash\{1\}} Z(\chi)$. Thus

$$\pi^{-1}(0) = \cap_{e\in\Lambda^1} D_e.$$

On the other hand, if $g \in \mathcal{M}$, we have observed above that $g|\pi^{-1}(0) \neq 0$. Hence, for any $e \in \Lambda^1$, $D_e \not\subseteq Z(g)$. But for any $f \in L(M)$,

$$Z(f) = (\cup_{\alpha\in C} D_\alpha) \cup (\cup_{e\in B} D_e)$$

(where $C \subseteq \Delta$ and $B \subseteq \Lambda^1$) since any $B \times B^-$-invariant prime divisor D of M is either a D_α or else a D_e. We conclude that

$$Z(g) \subseteq \cup_{\alpha\in\Delta} D_\alpha.$$

e) implies b). If we have such a factorization $f = \chi g$, for any $f \in L(M)$, we need to show two things:

i) $\mathcal{M} = \{g \in L(M) \mid Z(g) \subseteq \cup D_\alpha\}$;
ii) the factorization $f = \chi g$, with $\chi \in X(Z)$ and $g \in \mathcal{M}$, is unique.

Clearly, $\{g \in L(M) \mid Z(g) \subseteq \cup D_\alpha\} \subseteq \mathcal{M}$, since if $Z(g) \subseteq \cup D_\alpha$ and $g = \chi h$, with $\chi \in X(\overline{Z})$, then $\chi \in K[M]$ is a unit (since otherwise, $D_e \subseteq Z(g)$ for some $e \in \Lambda^1$). Hence $\chi = 1$, since M has a zero element. Conversely, if $f \notin \{g \mid Z(g) \subseteq \cup D_\alpha\}$, then by assumption we can write $f = \chi g$, where $\chi \in X(\overline{Z})$ and $Z(g) \subseteq \cup D_\alpha$. Clearly, $f > g$, so that f is not a minimal element of $L(M)$. This proves i). To prove ii), assume that $\chi_1 g_1 = \chi_2 g_2$, with $\chi_i \in X(Z)$ and $g_i \in \mathcal{M}$. Then $Z(\chi_i) \subseteq \cup D_e$, while $Z(g_i) \subseteq \cup D_\alpha$. Hence $Z(\chi_1) = Z(\chi_2)$ and $Z(g_1) = Z(g_2)$. All the zeros and poles of $\chi_1\chi_2^{-1}$ are in $\{D_e\}$, while all the zeros and poles of $g_2 g_1^{-1}$ are in $\{D_\alpha\}$. Yet $\chi_1\chi_2^{-1} = g_2 g_1^{-1}$. Thus, $\chi_1\chi_2^{-1}$ has neither zeros nor poles. Hence $\chi_1 = \chi_2$.

e) implies c). Given $\lambda = \sum_{\alpha\in\Delta} c_\alpha\lambda_\alpha \in X(T_0)_+$, we have the irreducible representation $\rho : G_0 \to Gl(V)$ with highest weight λ. Furthermore, by Theorem 5.3 of [42], there is a unique $g_\lambda \in L(G_0)$ (see Definition 2.50) such that

$\nu_\alpha(g_\lambda) = c_\alpha$ for each $\alpha \in \Delta$, where ν_α is the valuation of $K[G_0]$ associated with D_α. On the other hand, there is a character $\chi : Z \to K^*$ such that

$$\chi \otimes \rho : Z \times G_0 \to Gl(V)$$

factors through

$$Z \times G_0 \to G \subseteq M.$$

Here, $Z \times G_0 \to G$ is the multiplication map. Thus $f = \chi g_\lambda \in L(M)$. By assumption, f factors as $f = \mu g$, where $Z(g) \subseteq \cup_{\alpha\in\Delta} D_\alpha$ and $\mu \in X(\overline{Z})$. But $\nu_\alpha(g) = c_\alpha$ for all $\alpha \in \Delta$. Hence $c(\lambda) = [Z(g)] = 0 \in Cl(M)$.

c) implies e). Let $f \in L(M)$. Then there is a representation $\rho : M \to End(V)$ such that $\rho|G_0$ has highest weight $\lambda = \sum_{\alpha\in\Delta} c_\alpha \lambda_\alpha$, where $\nu_\alpha(f) = c_\alpha$ for each $\alpha \in \Delta$. By the assumption of c), we can find $g \in L(M)$ such that $\nu_\alpha(g) = c_\alpha$ for each α, and $\nu_D(g) = 0$ for all prime divisors $D \subseteq M\backslash G$. Hence we let $\chi = fg^{-1}$. Then $\nu_\alpha(\chi) = 0$ for all $\alpha \in \Delta$, while $\nu_D(\chi) = \nu_D(f) \geq 0$ for any $D \subseteq M\backslash G$. Thus $\chi \in X(\overline{Z})$, and $f = \chi g$ is the desired factorization.

The proof that d) and e) are equivalent is left to the reader.

Definition 6.11. *A reductive monoid M is called* flat *if the conditions of Theorem 6.10 are satisfied.*

Corollary 6.12. *Let M be flat. Then*

$$\mathcal{M} = \{f \in L(M) \mid Z(f) \subseteq \cup_{\alpha\in\Delta}\overline{Bs_\alpha B^-}\}.$$

Corollary 6.13. *If $Cl(M) = \{0\}$ then M is flat.*

Example 6.14. $M_n(K)$ is flat. One can use Corollary 6.13 above. But one can also show this directly using the calculations of Example 5.22. Recall from that example the "determinant functions"

$$\varphi_e(x) = \det(exe + 1 - e).$$

Then $\mathcal{M} =< \varphi_{e_1}, \ldots, \varphi_{e_{n-1}} >$, the submonoid of $K[M_n(K)]$ generated by the φ_{e_i}.

Let M be a reductive normal monoid with abelization $\pi : M \to A$. The following theorem was also obtained by Vinberg in [142].

Theorem 6.15. *The following are equivalent.*

a) M is flat.
b) There exists a morphism $\theta : Z \to T_0$ of algebraic groups such that
  i) $\theta|Z_0$ is the identity;
  ii) $L(M) = \{(\chi, f_\lambda(0)) \in L(Z \times G_0) \mid \chi\theta^*(\lambda)^{-1} \in X(A)\}$.

Proof. Assume that M is flat, and let $f \in \mathcal{M} \subseteq L(M)$. Then we can write $f = f_\lambda$, where $\lambda \in L(G_0) \cong X(T_0)_+$. Indeed, $\mathcal{M}$ is identified with $X(T_0)_+$ via

$$\mathcal{M} \subseteq L(M) \subseteq L(Z \times G_0) \to L(G_0) \cong X(T_0)_+.$$

Hence we define $\theta^*(\lambda) = f_\lambda|\overline{Z} \in X(\overline{Z})$. Notice that $\theta^*(\lambda)|Z_0 = f_\lambda|Z_0$ for all $\lambda \in L(G_0)$. Now $f_\lambda = \theta^*(\lambda)f_\lambda(0) \in L(Z \times G_0)$, where $f_\lambda(0) = f_\lambda|G_0$. Notice also that θ^* extends uniquely to a homomorphism $\theta : X(T_0) \to L(\overline{Z}) = X(\overline{Z})$.

Suppose that $\chi g_\mu(0) \in L(Z_0 \times G_0)$ is such that $\chi\theta^*(\mu)^{-1} \in X(A)$. Then $\chi g_\mu(0) = (\chi\theta^*(\mu)^{-1})(\theta^*(\mu)f_\mu(0)) \in L(M)$, since $\chi\theta^*(\mu)^{-1} \in X(A)$ and $\theta^*(\mu)f_\mu(0) \in \mathcal{M}$. Conversely, if $\chi f_\mu(0) \in L(M) \subseteq L(Z \times G_0)$, then by our assumption $\chi f_\mu(0) = \delta f_\mu$, where $\delta \in X(A)$ and $f_\mu \in \mathcal{M}$. Furthermore, the decomposition is unique. But $f_\mu = \theta^*(\mu)f_\mu(0)$, and so $\chi f_\mu(0) = \delta f_\mu = \delta\theta^*(\mu)f_\mu(0)$. Thus $\chi = \delta\theta^*(\mu)$. Hence $\chi\theta^*(\mu)^{-1} = \delta \in X(A)$.

We have shown that, if M is flat, then $L(M) = \{(\chi, f_\lambda(0)) \in L(Z \times G_0) \mid \chi\theta^*(\lambda)^{-1} \in X(A)\}$. Furthermore, θ^* satisfies property b)i).

Now assume b), so that there exists $\theta^* : X(T_0) \to X(Z)$ such that $f_\lambda(0)|Z_0 = \theta^*(\lambda)|Z_0$ for all λ, and $L(M) = \{(\chi, f_\lambda(0)) \in L(Z\times G_0) \mid \chi\theta^*(\lambda)^{-1} \in X(A)\}$. Then for $(\chi, f_\lambda(0)) \in L(M)$, we can write

$$(\chi, f_\lambda(0)) = (\chi\theta^*(\lambda)^{-1}, 1)(\theta^*(\lambda), f_\lambda(0))$$

so that

$$\mathcal{M} = \{(\theta^*(\lambda), f_\lambda(0)) \in L(Z \times G_0) \mid f_\lambda(0) \in L(G_0)\}.$$

Thus, condition b) of Theorem 6.10 is satisfied and, consequently, M is flat.

It turns out that that there is a universal, flat monoid $Env(G_0)$, associated with each semisimple group G_0. This amazing monoid was originally discovered and constructed by Vinberg in [142]. He refers to it as the **enveloping semigroup** of G_0. It has the following universal property:

Let M be any flat monoid with zero. Assume that the semisimple part of the unit group of M is G_0. Let $A(M)$ denote the abelization of M, as in Theorem 6.3, and let $\pi_M : M \to A(M)$ be the abelization morphism. (We make one exception with this notation. We let A denote the abelization of $Env(G_0)$. Let $\pi : Env(G_0) \to A$ be the abelization morphism.)

Given any isomorphism φ_0 from the semisimple part of $G(M)$ to the semisimple part of $Env(G_0)$, there are unique morphisms

$$a : A(M) \to A$$

and

$$\varphi : M \to Env(G_0)$$

such that

i) $\varphi|G_0 = \varphi_0$;
ii) $a \circ \pi_M = \pi \circ \varphi$;
iii) $\phi : M \cong E(a, \pi)$, via $\phi(x) = (\pi_M(x), \varphi(x))$, where $E(a, \pi) = \{(x, y) \in A(M) \times Env(G_0) \mid a(x) = \pi(y)\}$, is the **fibred product** of $A(M)$ and $Env(G_0)$ over A.

There are several ways to construct this monoid $Env(G_0)$, and there are already hints in Theorem 6.10. However, we use the construction in Theorem 17 of Rittatore's thesis [120]. The reader should also see Vinberg's construction in [142]. Notice that we are using multiplicative notation. In particular, $X(T_0/Z_0)_+$ is the the subgroup of $X(T_0)$ generated by the positive roots.

Theorem 6.16. *Let G_0 be a semisimple group and let*

$$\mathfrak{L}(G_0) = \{(\chi, \lambda) \in L(T_0 \times G_0) \mid \chi\lambda^{-1} \in X(T_0/Z_0)_+\}.$$

Define

$$K[Env(G_0)] = \bigoplus_{(\chi,\lambda)\in\mathfrak{L}(G_0)} (V_\lambda \otimes V_\lambda^*) \otimes \chi \subseteq K[G_0 \times T_0].$$

Then $K[Env(G_0)]$ is the coordinate algebra of the normal, reductive algebraic monoid $Env(G_0)$ with the above-mentioned universal property. In particular, $L(Env(G_0)) = \mathfrak{L}(G_0)$.

Proof. It follows from Vinberg's criterion in Theorem 5.10 that $K[Env(G_0)]$ is the coordinate algebra of a normal, reductive monoid. Indeed, this follows directly from the defining conditions of $L(G_0)$, taking into account the fact that, for all $\lambda, \mu \in X_+$, any $\nu \in X(\lambda, \mu)$ has the form

$$\nu = \lambda + \mu - \sum k_i \alpha_i$$

where $k_i \geq 0$. Also,

$$G(A) \cong T_0/Z_0$$

via $K[A] = \bigoplus_{(\chi,f_\lambda)\in X(A)} V_\lambda \otimes V_\lambda^* \otimes \chi$, where

$$X(A) = \{(\chi, f_\lambda) | dim(V_\lambda) = 1\} \cong X(T_0/Z_0)_+.$$

$Env(G_0)$ is flat by the criterion of Theorem 6.15. Indeed, if $(\chi, f_\lambda) \in L(Env(G_0))$, we can write

$$(\chi, f_\lambda) = (\chi\lambda^{-1}, 1)(\lambda, f_\lambda)$$

with $(\chi\lambda^{-1}, 1) \in X(A) \subseteq L(M)$, and $(\lambda, f_\lambda) \in L(M)$. Hence

$$\mathcal{M} = \{(\lambda, f_\lambda) \mid \lambda \in X(T_0) = X(\overline{Z}\}.$$

With these identifications, $\theta^* : X(T_0)_+ \to X(Z) = X(T_0)$ is just the inclusion. In particular, θ^* extends to an isomorphism $\theta^* : X(T_0) \to X(Z)$.

Now let M be any flat reductive monoid with semisimple part G_0 and zero element $0 \in M$. We assume that the semisimple part G_0 of the unit group of M has been identified with $G_0 \subseteq Env(G_0)$.

So $L(M) = \{(\chi, f_\lambda(0)) \in L(Z \times G_0) \mid \chi\theta^*_M(\lambda)^{-1} \in X(A)\ \}$, where $\theta^*_M : X(T_0) \to X(Z)$. Define $a^* : X(Z(Env(G_0))) \to X(Z(M))$ by

$$a^*(\chi) = \theta^*_M(\gamma(\chi)),$$

where γ is the inverse of θ^*, and define $\varphi^* : L(Env(G_0)) \to L(Z \times G_0)$ by

$$\varphi^*(\chi, f_\lambda(0)) = (\theta^*_M(\gamma(\chi)), f_\lambda(0)).$$

It can be checked, as in the proof of Theorem 5 of [142], that $\varphi^*(L(Env(G_0)) \subseteq L(M) \subseteq L(Z \times G_0)$, and consequently $\varphi : G \to Env(G_0)$ extends to a morphism $\varphi : M \to Env(G_0)$. It then follows from the definition of $L(M)$, along with a diagram chase, that $L(M)$ is the result of a pushout of $a^* : X(A) \to X(A(M))$ and $\pi^* : X(A) \to L(Env(G_0))$. This kind of pushout turns into a tensor product over $K[A]$ on the level of coordinate algebras. Thus,

$$K[M] \cong K[A(M)] \otimes_{K[A]} K[Env(G_0)],$$

which is the coordinate ring of the sought-after fibred product. The morphisms φ and a are unique because $Z \cdot G_0$ is dense in M.

Example 6.17. Let $M = \{(x, y, z) \in M_2(K) \times K^2 \mid det(x) = yz\}$. In this case, $G_0 = Sl_2(K)$, and so $Env(G_0) = M_2(K)$. Also $A(M) = \{(d, y, z) \in K^3 \mid d = yz\} \cong K^2$. Thus,

$$M \cong E(a, \pi)$$

where $\pi : M_2(K) \to K$ is the determinant, and $a : A(M) \to K$ is given by $a(d, y, z) = d = yz$

It is useful to know the type map of $Env(G_0)$ (see Definition 4.6). Recall that this is essentially a description of the $G \times G$-orbits of $Env(G_0)$, along with enough information to build the monoid from these orbits. The lattice of orbits was calculated by Vinberg in [142]. We describe his results in a way that allows us to relate $Env(G_0)$ to certain other monoids associated with G_0. Our proof here is somewhat sketchy. See [142] for more details.

Theorem 6.18. *Let Λ denote the cross section lattice of $Env(G_0)$. Then $\Lambda = \{e_{I,X} \mid I, X \subseteq S$, and no component of X is contained in $S \backslash I\}$. Λ is ordered as follows:*

$$e_{I,X} \geq e_{J,Y} \text{ if and only if } I \subseteq J \text{ and } X \subseteq Y.$$

Furthermore, the type map of $Env(G_0)$ is given as follows:

$\lambda^*(e_{I,X}) = \{s \in S\backslash I \mid sx = xs \text{ for all } x \in X\}$,
$\lambda_*(e_{I,X}) = X$.

Furthermore, $\lambda^1(e_{I,X}) = \{e_\alpha \mid \alpha \in I\}$, *where* e_α *is short for* $e_{\alpha,\phi}$.

Proof. Notice that $Cl(Env(G_0))$ is a finite group. Then for each $e \in \Lambda^1$ there is a $\chi \in X(A)$ such that $\chi^{-1}(0) = MeM$. Thus, $\Lambda^1 = \{e_\alpha \mid \alpha \in S\}$, since A is a simplicial, affine torus embedding with zero element. Then each convergent $1 - PSG$, $\lambda : K^* \to A$, has limit $f = f_I = lim_{t \to 0}\lambda(t) \in E(A)$. Also $I \subseteq S$ equals $\{\alpha \in S \mid \pi(e_\alpha) \geq f\}$. Now $C = \lambda(K) \subseteq A$ is one-dimensional, so the inverse image

$$M_I = \lambda^{-1}(C)$$

is a semisimple flat monoid. Thus, $M_I \backslash G(M_I)$ is an irreducible, algebraic variety (such monoids are called $\mathcal{J}$-coirreducible since $|\Lambda^1| = 1$). It is known [97] that the cross section lattice of M_I is

$$\Lambda_I = \{X \subseteq S \mid \text{no component of } X \text{ is contained in } S\backslash I\} \sqcup \{1\}.$$

Furthermore, for $e_X \in \Lambda_I$, $\lambda_*(e_X) = X$. Hence $\lambda^*(e_X) = \{s \in S\backslash I \mid sx = xs \text{ for all } x \in X\}$.

If we denote $e_X \in \Lambda_I$ by $e_{I,X}$, we see that

$$\Lambda = \{1\} \sqcup_{I \in S} \Lambda'_I$$

where $\Lambda'_I = \Lambda_I \backslash \{1\}$. Putting these all together yields the desired results.

A peculiar yet intriguing by-product of $Env(G_0)$ is the irreducible algebraic semigroup, $\pi^{-1}(0) \subseteq Env(G_0)$. Vinberg [143] calls it the **asymptotic semigroup** of G_0 since, like the asymptotic cone of a hyperboloid, it canonically reflects the behaviour of G_0 at infinity. He denotes this semigroup by $As(G_0)$. His construction also depicts G_0 as a flat deformation of $As(G_0)$.

See Theorem 10.19 for more information about $\mathcal{J}$-irreducible monoids.

From the proof of Theorem 6.18 we see that $As(G_0) = M_S \backslash G(M_S)$, where M_S is any $\mathcal{J}$-coirreducible monoid of type S. This means that if $\Lambda^1 = \{e_\phi\}$, then $\lambda^*(e_\phi) = \phi = S\backslash S$. Thus if $e_\phi \in \overline{T}$, then $e_\phi\overline{T}$ is a simplicial affine torus embedding with zero. Furthermore, the distinct idempotents of $e_\phi\overline{T}$ are contained in distinct $G \times G$-orbits of $As(G_0)$. Hence there are exactly 2^r such orbits, where r is the semisimple rank of G. The cross section lattice of any M_S is

$$\Lambda = \{1\} \sqcup \{e_I \mid I \subseteq S\}.$$

The type map λ of M_S is determined by λ_*. Furthermore

$$\lambda_*(e_I) = I.$$

It is likely that all the results of this section could be extended to the case of positive characteristics. Rittatore has already made an important contribution in this direction in [123].

6.4 Multilined Closure

Let M be a reductive monoid with zero. Associated with each minimal nonzero $\mathcal{J}$-class $J \in \mathcal{U}(M)$, there is an irreducible representation $\rho : M \to End(V)$ such that $\rho(e) \neq 0$ for any $e \in E(J)$. So let $\{J_1, \ldots, J_m\}$ be the minimal nonzero $\mathcal{J}$-classes of M, and let ρ_i, $i = 1, \ldots, m$, be the corresponding irreducible reresentations of M such that $\rho_i(J_i) \neq 0$. Assume that $M_j = \overline{\rho_j(M)}$ is a **$\mathcal{J}$-irreducible monoid of type** I_j. By this we mean that M_j is $\mathcal{J}$-irreducible and, for any minimal nonzero idempotent e of M_j, $P_\lambda(e)$ is a parabolic subgroup of $G(M_j)$ of type I_j. See § 7.3 for a detailed discussion of $\mathcal{J}$-irreducible monoids. In particular, notice that the cross section lattice of a $\mathcal{J}$-irreducible monoid may be identified with a subset of $\mathcal{P}(S)$, the set of subsets of the set of simple roots. In fact, the cross section lattice Λ_I of the $\mathcal{J}$-irreducible monoid M_I of type I is

$$\Lambda_I = \{e_A \mid A \subseteq S, \text{ and no component of } A \text{ is contained in } I\} \cup \{0\}.$$

Then if $e \in \Lambda_I$, we have either $e = e_A$ for some $A \subseteq S$, or else $e = e_0 = 0$. But the "zero" here is not a subset of S. Let

$$\rho = (\rho_1, \ldots, \rho_m) : M \to \Pi End(V_i).$$

Definition 6.19. *We define a* multilined closure of type $\mathbf{I}$=$(I_1, \ldots, I_m)$ *to be the closure* $M(I_1, \ldots, I_m)$ *of* $\rho(M)(K^* \times .... \times K^*)$ *in* $\Pi End(V_i)$. *Then* $M(I_1, \ldots, I_m)$ *is called* the multilined closure associated with M.

The multilined closure was first discussed in [52]. The following structure theorem was obtained.

Theorem 6.20. *Let Λ_i (respectively, λ_i) be the cross section lattice (respectively, type map) of M_i. Define the following subset of $\Lambda_1 \times \cdots \times \Lambda_m$:*

$$\Lambda_\mathbf{I} = \{(e_{Y_1}, \ldots, e_{Y_m}) | Y_i \subseteq \lambda_j(e_{Y_j}), \textit{ whenever } Y_i \neq 0\}.$$

Define

$$\lambda_\mathbf{I}(e_{Y_1}, \ldots, e_{Y_m}) = \cap_{i=1}^m \lambda_i(e_{Y_i}).$$

If Λ is the cross section lattice of $M(\mathbf{I})$ and λ is its type map, then

$\Lambda = \Lambda_\mathbf{I}$, *and*
$\lambda = \lambda_\mathbf{I}$.

The natural map $M \to M(\mathbf{I})$ induces a bijection on minimal nonzero $\mathcal{J}$-classes. In particular, it is a finite morphism.

Proof. By the results of § 7.3, $\Lambda \subseteq \Lambda_1 \times \cdots \times \Lambda_m$. Let $(e_1, \ldots, e_m) \in \Lambda$, $e_i = e_{Y_i}$, $i = 1, \ldots, m$. We claim that $e \in \Lambda_{\mathbf{I}}$. Otherwise, for some nonempty sets $Y_i$and Y_j, we would obtain $Y_i \not\subseteq Y_j$. Then for some simple root $\alpha \in \Delta$, $s = s_\alpha \in Y_i$ while $s \notin \lambda_j(Y_j)$. Let $P = P(e)$ and $P^- = P^-(e)$. Then the root subgroup of α,

$$U_\alpha \subseteq R_u(P_{Y_j}) \subseteq R_u(P),$$

since $P \subseteq P_{Y_j}$. Hence $U_\alpha e = e$. Since $s \in C_W(e_i)$, $U_\alpha \subseteq C_G(e_i)$. Let

$$H = \{g \in G \mid ge_i = e_i g = e_i\} \subseteq C_G(e_i).$$

Since $H \subseteq C_G(e_i)$ is a normal subgroup with $X_\alpha \subseteq H$, $s \in W(H)$, the Weyl group of H. Now $Q = C_B(e_i)H$ is a parabolic subgroup of $C_G(e_i)$, and thus $Q = C_B(e_i)W_K C_B(e_i)$. Since H is normal in $C_G(e_i)$, the component $A \subseteq S$ of s in Y_i is contained in K. Let $\{f_i\}$ be the set of minimal elements of Λ_i. Then $e_i \geq f_i$, and

$$Qf_i \subseteq HBf_i = Hf_iBf_i = f_iBf_i.$$

Hence $Q \subseteq P(f_i) = P_{I_i}$, and so $A \subseteq I_i$, a contradiction. Hence $e \in \Lambda_{\mathbf{I}}$. Clearly, $\lambda(e) = \lambda_{\mathbf{I}}(e)$.

We now prove the converse, namely that $\lambda_{\mathbf{I}}(e) \subseteq \lambda$. For $Y \subseteq S$, let

$$e_Y = (e_{Y_1}, \ldots, e_{Y_m})$$

where Y_i is the union of the components of Y not contained in I_i. Then for all i, j,

$$Y_i \subseteq Y \subseteq \lambda_j(e_j).$$

Hence $e_Y \in \Lambda_{\mathbf{I}}$. Clearly, $e_Y e_Z = e_{Y \cap Z}$, for all $Y, Z \subseteq S$. Let

$$\widehat{Y} = \{(e_{Z_1}, \ldots, e_{Z_m}) \mid Z_i = 0 \text{ or } Y_i\}.$$

We claim that

$$\Lambda_{\mathbf{I}} = \cup_{Y \subseteq S} \widehat{Y}.$$

Let $0 \neq e = (e_1, \ldots, e_m) \in \Lambda_{\mathbf{I}}$, $Y = \lambda(e)$. If $e_i = e_Z \neq 0$, then

$$Z \subseteq Y \subseteq \lambda_i(e_i).$$

Hence, $Y_i = Z$, and so $e \in \widehat{Y}$. This proves the claim.

To finish the proof, we need to show that $\Lambda_{\mathbf{I}} \subseteq \Lambda$, and we do this by induction on m. If $m = 1$, this is just the $\mathcal{J}$-irreducible case. So assume that $m > 1$. By the above discussion, if $(e_1, \ldots, e_m) \in \Lambda$ and $e_i = 1$, then

$$e_j = 1 \text{ or } 0 \text{ for all } j = 1, \ldots, m. \tag{$*$}$$

Let P be a standard maximal parabolic subgroup of G. Let P^- be its opposite, and let $L = P \cap P^-$. Let T_0 be the identity component of the center of L. Then $\dim(T_0)$=1. For $i = 1, \ldots, m$, let

$$E(\overline{\{\alpha\rho_i(t) \mid t \in T_0\}}) = \{1, e_i, f_i, 0\}$$

with $P(e_i) = P^-(f_i) = P$. By the above discussion (concerning $\widehat{Y}$), it suffices to show that $(e_1, \ldots, e_m) \in \Lambda$. But, by the induction hypothesis, $(e_1, \ldots, e_{m-1}, 0) \in \overline{E(T_1)}$, where

$$T_1 = \{(\alpha_1\rho_1(t), \ldots, \alpha_m\rho_m(t)) \mid \alpha_i \in K^*,\ i = 1, \ldots, m\}$$

Clearly,

$$E(\overline{T_1}) \subseteq \oplus_{i=1}^m \{1, e_i, f_i, 0\}.$$

By $(*)$, $1 = (1, \ldots, 1)$ covers $(1, \ldots, 1, 0)$, and $(1, \ldots, 1, 0)$ covers $(e_1, \ldots, e_{m-1}, 0)$ in $E(\overline{T_1})$. But $E(\overline{T_1})$ is a relatively complemented lattice. Hence there exists $e \in E(T_1)$ such that $e \neq (1, 1, \ldots, 1, 0)$, and such that

$$(e_1, \ldots, e_{m-1}, 0) < e < 1$$

So again by $(*)$, either $e = (e_1, \ldots, e_{m-1}, e_m)$, or else $e = (e_1, \ldots, e_{m-1}, f_m)$. However, $P(e_1, \ldots, e_{m-1}, f_m)$ is not parabolic. Hence $e = (e_1, \ldots, e_{m-1}, e_m)$, completing the proof.

Remark 6.21. This multilined closure construction behaves as if it is, in some sense, dual to the construction in Theorem 6.7. In the construction of M' in Theorem 6.7 the maximal idempotents have a special property, while in the multilined closure construction the minimal idempotents have a special property.

Example 6.22. Let $G_0 = Sl_3(K)$ so that $S = \{s_1, s_2\}$. Let $\rho_1 = id : G_0 \longrightarrow Gl_3(K)$, and let $\rho_2 : Sl_3(K) \longrightarrow Gl_3(K)$ be defined by $\rho_2(x) = (x^{-1})^t$. Define $M_1 = \overline{K^*\rho_1(G_0)}$, and $M_2 = \overline{K^*\rho_2(G_0)}$. Thus

$$M_1 \text{ is of type } J_1 = \{s_1\},$$

$$M_2 \text{ is of type } J_2 = \{s_2\}.$$

By Theorem 6.20, $M(J_1, J_2)$ has cross section lattice as depicted in the diagram below.

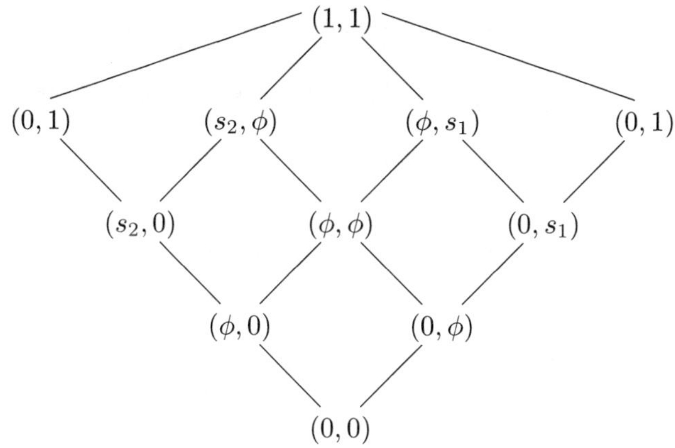

6.5 Normalization and Representations

As we have pointed out in § 3.2, any irreducible algebraic monoid M has a **normalization** $\eta : M' \to M$. Here M' has the unique structure of a normal, algebraic monoid such that η is a finite, birational morphism of algebraic monoids. Furthermore, η has the appropriate universal property.

Let K be an algebraically closed field of characteristic zero and let G be a semisimple group, defined over K. In §3 of [20], DeConcini provides a very revealing construction of the normalization of M_V, where M_V is the **lined closure** of the rational representation $\rho : G \to End(V)$ of the semisimple group G. A lined closure is the special case of the mutiltilined closure (see Definition 6.19), with $m = 1$. So, in our case,

$$M_V = \overline{\rho(G)K^*} \subseteq End(V)$$

where $K^* \subseteq End(V)$ is the set of nonzero homotheties.

Let V_λ be the irreducible representation of G with highest weight λ. Let Σ_λ be the **saturation** of λ. Thus

$$\Sigma_\lambda = \{\mu \mid \mu \text{ is dominant and } \mu \leq \lambda\}.$$

Along with DeConcini [20], we set

$$\mathcal{M}_\lambda = M_W,$$

where

$$W = \oplus_{\mu \in \Sigma_\lambda} V_\mu.$$

An irreducible representation V_λ is called **miniscule** if $\Sigma_\lambda = \{\lambda\}$. In this case λ is called a **miniscule weight**.

The following result is obtained in [20] (see his Theorem 3.1).

Theorem 6.23. *1. $\mathcal{M}_\lambda$ is a normal variety with rational singularities.*
2. $\mathcal{M}_\lambda$ is the normalization of M_λ.
3. $\mathcal{M}_\lambda$ and M_λ are equal if and only if λ is a miniscule weight.

His proof requires an application of the results of [21]. The proof also requires the result [45] of Kannan on the projective normality of the wonderful compactification.

Miniscule weights are the role models for the standard monomial theory. If λ is miniscule, the calculation of $V_{n\lambda} = H^0(G/P, L_{n\lambda})$ involves a striking blend of combinatorics and intersection theory. Hodge worked this out for $Sl_n(K)$, and then Seshadri extended Hodge's work to the case of any miniscule weight. See [37] for a good introduction.

6.6 Exercises

6.6.1 Flat Monoids

1. Let $f \in L(M)$. Prove that $Z(f) \subseteq (\cup_{\alpha \in \Delta} D_\alpha) \cup (\cup_{e \in \Lambda^1} D_e)$.
2. Let M be reductive. Show that $L(M)$ is a commutative, totally cancellative semigroup that embeds in a free abelian group of rank less than or equal to $|\Delta| + |\Lambda^1|$.
3. Prove the equivalence of d) and e) in Theorem 6.10.

7

Orbit Structure of Reductive Monoids

Let M be a reductive monoid with unit group G. We assume that M has a zero element $0 \in M$. The general case is not interestingly different (see Proposition 8 of [121]). From Theorem 4.2, M is regular, so that

$$M = GE(M) = E(M)G\ .$$

But we can do much better than this. Indeed, from Theorem 4.5,

$$M = \bigsqcup_{e \in \Lambda} GeG$$

where $\Lambda = \{e \in E(\overline{T}) \mid Be \subseteq eB\}$.

In this chapter we want to explain how M is "stuck together" using G, Λ and $P(e) = \{g \in G \mid ge = ege\}$. Since $P(e)$ is a parabolic subgroup containing B, the reader should take note of the key objective here: to obtain control of the structure of M in terms of something easily described in terms of the Coxeter-Dynkin complex of G, and the set of standard parabolic subgroups of G.

Our second objective here is to identify and record a large number of examples where we can determine Λ and $\Lambda \longrightarrow \mathcal{P}$, $e \rightsquigarrow P(e)$, explicitly. Notice that there is a canonical identification $\mathcal{P} = 2^S$, where S is the set of simple reflections. See Theorem 2.46. Thus we usually write the type map as $\lambda : \Lambda \longrightarrow 2^S$

7.1 The System of Idempotents and the Type Map

In this section we describe the orbit structure of a reductive monoid, assuming there is only one minimal, nonzero orbit. The results of this section are taken from [95].

Let M be reductive with unit group G, Borel subgroup $B \subseteq G$ and maximal torus $T \subseteq B$. $W = N_G(T)/T$. From Definition 4.6 we obtain the type map

$$\lambda : \Lambda \longrightarrow 2^S .$$

Recall that $\lambda(e) = \{s \in S \mid se = es\}$, where $S \subseteq W$ is the set of simple reflections of W relative to B, and Λ is the cross section lattice of M relative to B and T.

Notice that $\lambda(e)$ determines $P(e) = \{g \in G \mid ge = ege\}$ since $P(e)$ is generated by B and $\lambda(e)$. Note also that $P(e)$ and $P^-(e) = \{g \in G \mid eg = ege\}$ are opposite parabolic subgroups.

Definition 7.1. *Let*

$$E(\lambda) = \left\{ (J,P,Q) \middle| \begin{array}{l} J \in G\backslash M/G, \quad P \text{ and } Q \text{ are opposite parabolics,} \\ P = gP(e)g^{-1} \text{ for some } g \in G, \text{ where } J \cap \Lambda = \{e\} \end{array} \right\} .$$

A quasi-ordering *on a set E is a relation $\leq$ on E that is transitive and reflexive.*

Theorem 7.2. *Both $E(M)$ and $E(\lambda)$ have canonically defined quasi-orderings $\leq_\ell$ and $\leq_r$. Define*

$$\psi : E(M) \longrightarrow E(\lambda)$$

by $\psi(e) = (GeG,\ P(e),\ P^-(e))$. Then ψ is an isomorphism of biordered sets. $\leq_r$ and $\leq_\ell$ are defined as follows.

On $E(M)$ define

$$\begin{array}{l} e \leq_r f \text{ if } fe = e. \\ e \leq_\ell f \text{ if } ef = e. \\ e \leq f \text{ if } ef = fe = e. \end{array}$$

On $E(\lambda)$ define

$(J_1, P, Q)\mathcal{R}(J_2, P', Q')$ if $J_1 = J_2$ and $P = P'$.
$(J_1, P, Q)\mathcal{L}(J_2, P', Q')$ if $J_1 = J_2$ and $Q = Q'$.
$(J_1, P, Q) \leq (J_2, P', Q')$ if $J_1 \leq J_2$ and there exist opposite Borel subgroups B and B^-, such that $B \subseteq P \cap P'$ and $B^- \subseteq Q \cap Q'$.

Then define on $E(\lambda)$

$$\leq_r = \mathcal{R} \circ \leq$$

and

$$\leq_\ell = \mathcal{L} \circ \leq .$$

Proof. Assume that $GeG = GfG$. Then one checks, as in Lemma 3.4 of [95], that

$$\circledast \begin{cases} eM = fM \text{ if and only if } P(e) = P(f) \quad \text{and} \\ Me = Mf \text{ if and only if } P^-(e) = P^-(f) . \end{cases}$$

Next we check that ψ is bijective. Let $(J, P, Q) \in E(\lambda)$. Then $P = P(e)$ for some $e \in E(J)$. Further, by the results of [83], P is opposite to $P^-(e)$.

So by standard results there exists $g \in P$ such that $g^{-1}P^{-}(e)g = Q$. Thus, $(J, P, Q) = \psi(g^{-1}eg)$. Conversely, if $\psi(e) = \psi(f)$, then by ⊛ above that $eM = fM$ and $Me = Mf$. But then $e = f$, by an elementary semigroup calculation.

For the remainder of the proof we refer the reader to Theorem 3.5 of [95]. Notice that, in view of ⊛ above, all that remains here is to check that $e \le f$ if and only if $\psi(e) \le \psi(f)$.

The moral of the story is that, once we know λ, we obtain E automatically. If we also know G, then we can reconstruct M up to a kind of "central extension" abstractly. It can be seen direcly that $E(M)$ is a **biordered set** in the sense of Nambooripad [64].

7.2 The Cross Section Lattice and the Weyl Chamber

Let M be a reductive algebraic monoid. For the results of this section, it is not necessary to impose any other restrictions. We need to show how the cross-section lattice Λ can be described in terms relating $X(\overline{T})$ and the set of dominant weights

$$X(T)_{+} = \{\chi \in X(T) \mid \Delta_{\alpha}(\chi) \ge 0 \quad \text{for all } \alpha \in \Delta\}$$

where $\Delta_{\alpha} : X(T) \longrightarrow \mathbb{Z}$ is defined by the equation

$$\chi - s_{\alpha}(\chi) = \Delta_{\alpha}(\chi)\alpha.$$

Let $e \in E(\overline{T})$. Then $\overline{eT}$ is also a D-monoid with unit group eT. So let $X(eT)$ denote the monoid of characters of $\overline{eT}$. Consider

$$\mu_e = \left\{ \chi \in X(\overline{eT}) \subseteq X(\overline{T}) \,\middle|\, \begin{matrix} \chi \neq 0 \ and \\ \chi_{\overline{eT}\setminus eT} = 0 \end{matrix} \right\}$$

where $X(\overline{eT}) \subseteq X(\overline{T})$ via the map $\overline{T} \to \overline{eT}$, $z \to ez$. One can easily check that

$$X(\overline{T})\backslash\{0\} = \bigsqcup_{e \in E(\overline{T})} \mu_e \, .$$

Let Δ be the set of simple roots of G relative to B and T. For $\alpha \in \Delta$ let U_{α} be the one dimensional, unipotent subgroup of B, normalized by T with weight α.

Lemma 7.3. *Let $\alpha \in \Delta$ and $e \in E(\overline{T})$. The following are equivalent:*

a) $U_{\alpha}e = eU_{\alpha}e$;
b) either $U_{\alpha}e = eU_{\alpha}$, or else $U_{\alpha}f = f$ for all $f \in E_1(\overline{eT})$.

Proof. In case $eU_\alpha \neq e$ and $U_\alpha e \neq e$, one obtains that $U_\alpha e = eU_\alpha e$. Otherwise, if $U_\alpha f = f$ for all $f \in E_1(e\overline{T})$ yet $U_\alpha e \neq eU_\alpha e$ (i.e. $U_\alpha e \neq eU_\alpha$ and $U_\alpha e \neq e$), then $eU_\alpha = e \neq U_\alpha e$ is the only other possibility. Hence $\sigma_\alpha e \sigma_\alpha \neq e$ and thus, $\sigma_\alpha f \sigma_\alpha \neq f$ for some $f \in E_1(e\overline{T})$. But then $fU_\alpha = feU_\alpha = fe = f$. Hence $U_\alpha f \neq f$, since $f\sigma_\alpha \neq \sigma_\alpha f$. Contradiction.

Lemma 7.4. *a) The following are equivalent:*
i) $U_\alpha e = eU_\alpha$ *(equivalently,* $s_\alpha e = es_\alpha$*);*
ii) $\Delta_\alpha(\chi) = 0$ *for some* $\chi \in \mu_e$.
b) The following are equivalent (assuming $U_\alpha e \neq eU_\alpha$*).*
i) $U_\alpha f = f$ *for all* $f \in E_1(e\overline{T})$*;*
ii) $\Delta_\alpha(\chi) > 0$ *for some* $\chi \in \mu_e$.

Proof. For a), first note that, for $\chi \in \mu_e$, $\Delta_\alpha(\chi) = 0$ iff $s_\alpha(\chi) = \chi$. But $s_\alpha(\mu_e) = \mu_{e'}$ where $e' = s_\alpha e s_\alpha$. Hence $s_\alpha(\mu_e) \cap \mu_e \neq \phi$ iff $s_\alpha e s_\alpha = e$. Accordingly, if $\chi \in \mu_e$ and $s_\alpha(\chi) = \chi$, then $\chi \in s_\alpha(\mu_e) \cap \mu_e$. Conversely, if $\chi_1 \in s_\alpha(\mu_e) \cap \mu_e \neq \phi$, then $\chi_1 s_\alpha(\chi_1) \in \mu_e$ and $\Delta_\alpha(\chi) = 0$.

For b), assume first that $U_\alpha f = f$ for all $f \in E_1(e\overline{T})$. Now $\overline{fT} \cong K$ as algebraic varieties. Hence there is a unique character $\chi_f \in X(\overline{T})$ such that $\mathcal{O}(\overline{fT}) = K[\chi_f]$. But from Lemma 3.6 of [109], $\Delta_\alpha(\chi_f) \geq 0$. Now

$$K[\overline{eT}] = K[\chi \mid \chi^n \in \langle \chi_{f_1}, \ldots, \chi_{f_2} \rangle \text{ for some } n > 0] \qquad (*)$$

where $\{f_i\}_{i=1}^s = E_1(e\overline{T})$. Now if $\Delta_\alpha(\chi_f) = 0$ for all $f \in E_1(e\overline{T})$, then $s_\alpha(\chi_f) = \chi_f$ for all $f \in E_1(e\overline{T})$. Hence by $(*)$, $s_\alpha(K[\overline{eT}]) = K[\overline{eT}]$ and so $s_\alpha e = es_\alpha$. Thus $U_\alpha e = eU_\alpha$, a contradiction. Thus $\Delta_\alpha(\chi_f) > 0$ for some $f \in E_1(e\overline{T})$. Hence $\Delta_\alpha(\chi) > 0$ for all $\chi \in \mu_e$. Conversely, suppose that $\Delta_\alpha(\chi_f) > 0$ for some $f \in E_1(e\overline{T})$. Consider

$$\chi = \chi_f^N \chi_{f_2} \cdot \ldots \cdot \chi_{f_s} \in \mu_e$$

where $N > 0$ and $E_1(e\overline{T}) = \{f, f_2, \ldots, f_s\}$. Then $\Delta_\alpha(\chi) < 0$ if $N >> 0$. This is a contradiction.

Theorem 7.5. *The following are equivalent for* $e \in E(\overline{T})\backslash\{0\}$.

a) $e \in \Lambda\backslash\{0\}$*;*
b) there exists $\chi \in \mu_e$ *such that* $\Delta_\alpha(\chi) \geq 0$ *for all* $\alpha \in \Delta$.

Proof. Now $e \in \Lambda' := \Lambda\backslash\{0\}$ if and only if for all $\alpha \in \Delta$ either $U_\alpha e = eU_\alpha$ or else $U_\alpha e \neq eU_\alpha$ and $U_\alpha f = f$ for all $f \in E_1(\overline{eT})$. By Lemma 7.4 this is equivalent to:

For each $\alpha \in \Delta$, either

$$\Delta_\alpha(\chi) = 0 \text{ for some } \chi \in \mu_e$$

or else

$$\Delta_\alpha(\chi) > 0 \text{ for all } \chi \in \mu_e \,.$$

Thus, $e \in \Lambda\backslash\{0\}$ if and only if for all $\alpha \in \Delta$ either

i) $\Delta_\alpha(\chi) = 0$ for some $\chi \in \mu_e$, or else
ii) $\Delta_\alpha(\chi) > 0$ for all $\chi \in \mu_e$.

Hence b) implies a).

Conversely, if $e \in \Lambda\backslash\{0\}$ then $\Delta = \Delta_1 \sqcup \Delta_2$, where

$$\Delta_1 = \{\alpha \in \Delta \mid s_\alpha e = e s_\alpha\}$$

and

$$\Delta_2 = \{\alpha \in \Delta \mid s_\alpha e \neq e s_\alpha\}.$$

Let $\chi_0 \in \mu_e$ and define

$$\chi = \prod_{w \in W_{\Delta_1}} w(\chi_0) \in \mu_e \,.$$

Then $\Delta_\alpha(\chi) = 0$ for all $\alpha \in \Delta$. But $\Delta_\alpha(\chi) > 0$ for all $\alpha \in \Delta_2$.

Theorem 7.5 has a very appealing geometric interpretation.

One can identify $E(\overline{T})$ with the face lattice $\mathcal{F}$ of the rational polyhedral cone $X(\overline{T}) \otimes \mathbb{Q}^+ \subseteq X(T) \otimes \mathbb{Q}$. Furthermore, $X(\overline{T}) \otimes \mathbb{Q}^+$ is W-invariant. We can think of $\mu_e \otimes \mathbb{Q}^+$ as the topological interior of $X(\overline{eT}) \otimes \mathbb{Q}^+ \in \mathcal{F}$. Theorem 7.5 says that

$$\Lambda = \left\{ e \in E(\overline{T}) \middle| \begin{array}{l} \text{the interior of } X(e\overline{T}) \otimes \mathbb{Q}^+ \text{ meets} \\ X(\overline{T})_+ \otimes \mathbb{Q}^+ \end{array} \right\} .$$

Clearly, $|Cl_W(e) \cap \Lambda| = 1$ for all $e \in E(\overline{T})$.

Recall that a reductive monoid M is *semisimple* if the center of G is one-dimensional and M has a zero element. In this case the zero element of M is in the closure $\overline{Z}$ of Z the one-dimensional connected center of M. As Z is contained in any maximal torus T of G, we have in particular that $\overline{Z} \subseteq \overline{T}$. Thus we obtain the induced (dual) map on the corresponding character monoids:

$$\gamma : X(\overline{T}) \to X(\overline{Z}) \cong \mathbb{N}.$$

This γ determines, on the associated rational polyhedral cones, a homomorphism

$$\zeta : X(\overline{T}) \otimes \mathbb{Q}^+ \to X(\overline{Z}) \otimes \mathbb{Q}^+ \cong \mathbb{Q}^+,$$

by setting $\zeta = \gamma \otimes 1$. For M semisimple we make the following definition.

Definition 7.6. *Let*

$$\mathcal{P} = \zeta^{-1}(1).$$

$\mathcal{P}$ *is the* polytope *of* M.

From the above results, $\mathcal{P}$ is W-invariant, and the face lattice $\mathcal{F}$ of $\mathcal{P}$ is canonically identified with $E(\overline{T})$. Furthermore, we can identify Λ as a subset of $\mathcal{F}$ using Theorem 7.5.

Example 7.7. Let $M = M_n(K)$, the semisimple monoid of $n \times n$ matrices over K. In this case $Z = \{\alpha I_n \mid \alpha \in K^* \}$, where I_n is the identity $n \times n$ matrix. If T is the D-group of invertible diagonal matrices then $\overline{T}$ is the set of diagonal matrices and $\zeta : X(\overline{T}) \otimes \mathbb{Q}^+ \to X(\overline{Z}) \otimes \mathbb{Q}^+$ is easily identified with the map

$$\rho : (\mathbb{Q}^+)^n \to \mathbb{Q}^+$$

defined by $\rho(s_1, ..., s_n) = \sum_i s_i$. The polytope here is

$$\mathcal{P} = \{(s_1, ..., s_n) \in (\mathbb{Q}^+)^n \mid \sum_i s_i = 1 \}.$$

The face lattice of $\mathcal{P}$ is easily identified with $E(\overline{T})\backslash\{0\}$. Notice that characters are written additively in this setup.

7.3 $\mathcal{J}$-irreducible Monoids

We start with a simple lemma to focus our discussion.

Lemma 7.8. *Let M be a reductive monoid with zero $0 \in M$. Let $\Lambda \subseteq E(\overline{T})$ be a cross section lattice. The following are equivalent.*

a) $\Lambda\backslash\{0\}$ has a unique minimal element e_0 (so that $e_0 f = e_0$ for all $f \in \Lambda\backslash 0$);
b) there exists a rational representation $\rho : M \longrightarrow End(V)$ such that
  i) V is irreducible over M.
  ii) ρ is a finite morphism.

Proof. Assume that $\rho : M \longrightarrow End(V)$ is as in b). Suppose that $e_1, e_2 \in \Lambda\backslash 0$ are minimal elements yet $e_1 \neq e_2$. Then $e_1 M e_2 = 0$. But $\rho(Me_1)V$ and $\rho(Me_2)V$ are M-submodules of V. Thus, $\rho(Me_1)V = \rho(Me_2)V = V$. But then $\rho(e_1)V = \rho(e_1)\rho(Me_2)V = \rho(e_1 M e_2)V = 0$, so that $\rho(e_1) = 0$. But this is impossible since ρ is a finite morphism.

Now assume that $\Lambda\backslash\{0\}$ has a unique minimal element e. Let $\rho : M \longrightarrow End(W)$ be a finite morphism of algebraic monoids [82]. If we replace W by $\overline{W} = \bigoplus_{i=1}^{n} W_i/W_{i-1}$, where $W_0 \subseteq W_1 \subseteq \cdots \subseteq W_n = W$ is a composition series of W, then $gr(\rho) : M \longrightarrow End(\overline{W})$ is also a finite morphism since, by regularity of M, $gr(\rho)^{-1}(0) = \{0\}$. So assume that $W = \bigoplus_{i=1}^{n} V_i$ where each V_i is irreducible. Now $\rho(e) \neq 0$, since ρ is finite. Say $\rho(e)(V_1) \neq 0$. Thus we let $\rho_1 = \rho|V_1$. Then $\rho : M \longrightarrow End(V_1)$ is the desired irreducible representation.

Definition 7.9. *Let M be as in Lemma 7.8. We say that M is* $\mathcal{J}$-irreducible.

The major purpose of this chapter is to determine the cross section lattices and the type maps of $\mathcal{J}$-irreducible reductive moniods. But first, let us notice what determines the cross section lattice Λ and the type map $\lambda : \Lambda \to 2^S$.

Given M as in Lemma 7.8, we observe several discrete invariants.

i) the *type* of the representation $\rho : M \longrightarrow End(V)$
ii) $\lambda(e_0) = J_0 = \{s \in S \mid se_0 = e_0 s\} \subseteq S$
iii) $\{g \in G \mid ge_0 = ege_0\} = P(e_0) < G$.

To define the *type* of ρ, let $B \subseteq G$ be a Borel subgroup and let $L \subseteq V$ be the line such that $\rho(B)L = L$. Then the *type of* ρ is the parabolic subgroup

$$P = \{g \in G \mid \rho(g)L = L\}\ .$$

These invariants all amount to the same thing. Indeed, $P = P(e_0) = \bigsqcup_{w \in W_{J_0}} BwB$ and $L = e_0(V)$. So our mission here is as follows.

> Determine the type map $\lambda : \Lambda \to 2^S$ in terms of $\lambda(e_0) = J_0 \subseteq S$ where $e_0 \in \Lambda \backslash \{0\}$ is the minimal element.

For the remainder of this section we assume that M is a $\mathcal{J}$-irreducible monoid.

Lemma 7.10. *Let $e, f \in E(M)$ be nonzero idempotents. Then $P(e) = P(f)$ if and only if $e\mathcal{R}f$.*

Proof. If $e\mathcal{R}f$, then $f = eg$ for some $g \in G$. Hence $fe = e$. If $x \in \mathcal{R}(e)$, then $xe = exe$ and so $xf = xeg = exeg = fexeg = fexf$. Hence $f(xf) = f(fexf) = fexf = xf$. Thus, $P(e) \subseteq P(f)$. By symmetry, $P(f) \subseteq P(e)$.

Conversely, assume that $P(e) = P(f)$. Assume that $e \in \overline{T} \subseteq P(e)$. Now there exists $g \in P(f)$ such that $f' = gfg^{-1} \in \overline{T}$. Then $f\mathcal{R}f'$ and $P(f) = P(f')$. So without loss of generality, $f = f'$. Now let $h \in E_1(\overline{T})$ be such that $he = eh = h$. Then there exists a cross section lattice Λ such that $e, h \in \Lambda$. Then

$$\begin{aligned} B &= \{g \in G \mid gh = hgh \text{ for all } h \in \Lambda\} \\ &\subseteq P(e) \end{aligned}$$

since $e \in \Lambda$. But $B \subseteq P(e) = P(f)$. Hence by definition,

$$f \in \{h \in E(\overline{T}) \mid gh = hgh \text{ for all } g \in B\} = \Lambda\ .$$

Since $h \in \Lambda$ is the unique, nonzero, minimal element of Λ, we have $fh = h$. But this is true for any $h \in E_1(\overline{eT}) = \{h_1, \ldots, h_s\}$, the set of minimal, nonzero idempotents of $\overline{eT}$. Thus $f = e$ since $e = h_1 \vee \cdots \vee h_s$. Similarly, $f = fe$. So $e = f$.

Lemma 7.10 is the crux of the matter. Indeed, let

$$\lambda : \Lambda\backslash\{0\} \longrightarrow 2^S ,$$

where $\lambda(e) = \{s \in S \mid se = es\}$, be the type map. Then $P(e) = P_{\lambda(e)}$. Hence by Lemma 7.10, λ is injective. Thus it remains to find $\lambda(\Lambda\backslash\{0\}) \subseteq 2^S$ and to recover the $\mathcal{J}$-ordering on Λ from this image.

Definition 7.11. *For $e \in \Lambda$ define*

$$\lambda^*(e) = \{s \in S \mid se = es \neq e\}$$
$$\lambda_*(e) = \{s \in S \mid se = es = e\}.$$

It is easy to check that $e \geq f$ implies that both $\lambda^*(f) \subseteq \lambda^*(e)$ and $\lambda_*(e) \subseteq \lambda_*(f)$. In particular, $\lambda_*(e) \subseteq \lambda_*(e_0) = J_0$. But we can do much more here. It turns out that we can characterize $\{I \in 2^S \mid I = \lambda^*(e)$ for some $e \in \Lambda\backslash 0\}$ and that $\lambda^*(e)$ determines $\lambda_*(e) \subseteq J_0$.

First recall the graph structure on S:

$$s \text{ and } t \text{ are joined by an edge if } st \neq ts.$$

Therefore we can talk about the **connected components** of any subset of S. The following theorem is the main result of [95].

Theorem 7.12. *a) The following are equivalent for $I \subseteq S$.*
i) $I = \lambda^*(e)$ for some $e \in \Lambda\backslash\{0\}$.
ii) No connected component of I lies entirely in J_0.
Furthermore, if $e \geq f$ then $\lambda^*(e) \supseteq \lambda^*(f)$
b) For any $e \in \Lambda\backslash\{0\}$, $\lambda_*(e) = \{s \in J_0\backslash\lambda^*(e) \mid st = ts$ for all $t \in \lambda^*(e)\}$.

Proof. For b) we refer the reader to the straightforward calculation of Lemma 4.10 of [95]. For a), first notice that $e \mapsto \lambda^*(e)$ is an injection $\Lambda\backslash\{0\} \longrightarrow 2^S$ since λ is injective and it is determined by λ^*. Furthermore, if $e \geq f$ then $eMe \supseteq fMf$, while $\lambda^*(e)$ is canonically identified with the simple reflections of eMe (and similarly for f). Hence $\lambda^*(e) \supseteq \lambda^*(f)$.

To see why i) and ii) are equivalent, we start with e_0 and notice that $\lambda^*(e_0) = \phi$; and then work our way "up". The key step is Theorem 4.13 of [95].

For $e \in \Lambda\backslash\{0\}$ there is a canonical bijection between $\{f \in \Lambda\backslash\{0\} \mid f$ covers $e\}$ and

$$\{s \in S \mid se \neq es\}. \qquad (*)$$

Hence f corresponds to the unique s with $\lambda^*(f) = \lambda^*(e) \cup \{s\}$.

To find f given s, consider W_I where $I = \lambda^*(e) \cup \{s\}$. Then it is easily checked that there is a minimal element $e' \in \Lambda^{W_I} = \{e \in \Lambda \mid we = ew$ for all $w \in W_I\}$ such that $e'e = ee' = e$ and $e' \neq e$. This gives us "a foot in the doorway" since $\lambda^*(e') \supseteq \mu(e) \cup \{s\}$. We can now find f by induction on $\dim M$ using the $\mathcal{J}$-irreducible monoid $e'Me'$.

Now, given $I \subseteq S$ as in a) ii), let

$$\begin{aligned} K_1 &= S \backslash J_0 \\ K_2 &= K_2 \cup \{s \in I \backslash K_1 \mid st \neq ts_1, \text{ some } t \in K_1\} \\ &\vdots \\ K_i &= K_{i-1} \cup \{s \in I \backslash K_{i-1} \mid st \neq ts, \text{ some } t \in K_{i-1}\} \\ &\vdots \end{aligned}$$

By definition $I = K_s$ for some $s > 0$. But from (*) applied repeatedly, there exists $e_i \in \Lambda \backslash \{0\}$ such that $\lambda^*(e_i) = K_i$.

Remark 7.13. a) Theorem 7.12 provides an algorithm for calculating Λ and $\lambda : \Lambda \longrightarrow 2^S$ for any $\mathcal{J}$-irreducible monoid M in terms of $J_0 = \lambda(e_0) = \lambda_*(e_0)$. In each case, $S \backslash J_0$ corresponds to the set of fundamental dominant weights involved in the associated irreducible representation of M.

b) One defines a reductive monoid M, with zero, to be $\mathcal{J}_i$-irreducible if $|\Lambda_j| = 1$ for all $j \leq i$. The reader can check that
   i) M is $\mathcal{J}_2$-irreducible if and only if $J_0 = S \backslash \{s\}$ for some $s \in S$,
   ii) M is $\mathcal{J}_3$-irreducible if and only if $J_0 = S \backslash \{s\}$ where s corresponds to an end node on the Dynkin diagram of G. See Figure 7.1 below.

c) One can use Theorem 7.12 to characterize other classes of $\mathcal{J}$-irreducible monoids.
   i) We say that a semisimple monoid M is **$\mathcal{J}$-simple** if each $\mathcal{H}$-class of M has at most one simple component. It turns out that M is $\mathcal{J}$-simple if and only if S is connected and M is either $\mathcal{J}_2$-irreducible or $S \backslash J_0 = \{s, t\}$ where $st \neq ts$. See Figure 7.2 below and Exercise 3 of 7.7.1.
   ii) $\Lambda(M)$ is a distributive lattice if and only if $S \backslash J_0$ is connected.

7.4 Explicit Calculations of the Type Map

In this section we illustrate Theorem 7.12 by using it to calculate the type maps of several interesting classes of $\mathcal{J}$-irreducible monoids. In our first example we calculate the type maps associated with the adjoint representation.

7.4.1 The Type Map for the Adjoint Representations

In this subsection we illustrate Theorem 7.12 by calculating the type maps for the monoids associated with the adjoint representations of simple groups. Here $M = \overline{K^* Ad(G)} \subseteq End(\mathfrak{g})$ where $\mathfrak{g}$ is the Lie algebra of G. Also, by b) of Theorem 7.12, it suffices to calculate

$$\{\lambda^*(e) \mid e \in \Lambda\backslash\{0\}\}$$

in each case. We also include the Hasse diagram of Λ in each case, along with an illustration of the corresponding extended Dynkin diagram. The reader can use the extended Dynkin diagram to "see" how Theorem 7.12 is used to calculate Λ.

a) **Type A_ℓ:**

$$\begin{aligned} &S = \{s_1, \ldots, s_\ell\} \\ &s_i s_j \neq s_j s_i \quad \text{if } |i-j| = 1 \\ &J_0 = \{s_2, \ldots, s_{\ell-1}\} \end{aligned}$$

$$\lambda^*(\Lambda\backslash\{0\}) = \{S\backslash\{s_i, s_{i+1}, \ldots, s_j\} \mid 1 \leq i \leq j \leq \ell\} \cup \{S\} .$$

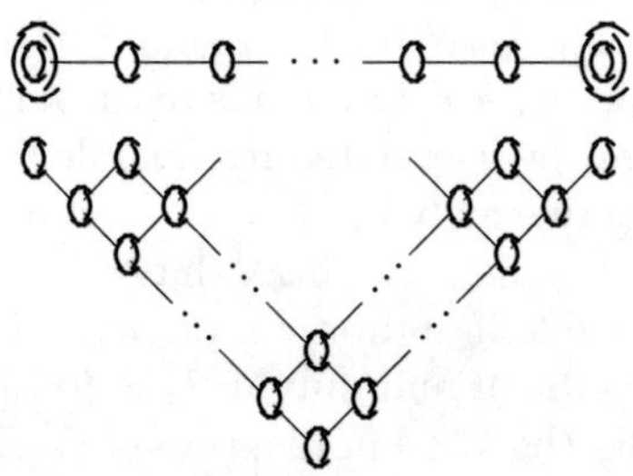

b) **Type B_ℓ:**

$$\begin{aligned} &S = \{s_1, \ldots, s_e\} \\ &s_i s_j \neq s_j s_i \quad \text{if } |i-j| = 1 \\ &J_0 = \{s_1, s_3, \ldots, s_\ell\} \end{aligned}$$

$$\begin{aligned} \lambda^*(\Lambda\backslash 0) = \{&\phi; \{s_2\}; \{s_1, s_2\}, \{s_2, s_3\}; \ldots \\ &\ldots; \{s_1, s_2, \ldots, s_i\}, \{s_2, s_3, \ldots, s_{i+1}\}; \ldots \\ &\ldots; \{s_1, \ldots, s_{\ell-1}\}, \{s_2, \ldots, s_\ell\}; \{s_1, \ldots, s_\ell\}\}. \end{aligned}$$

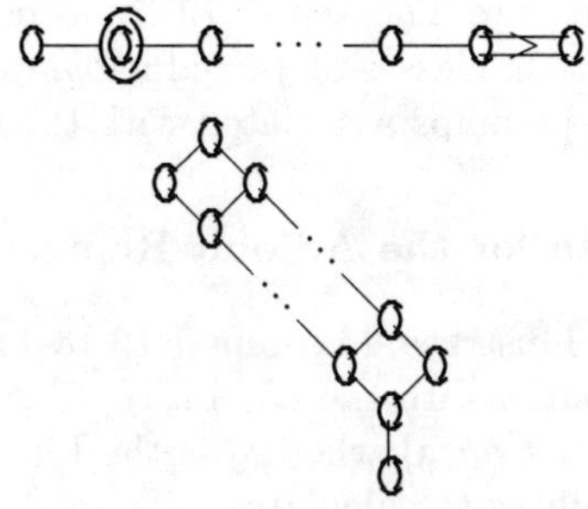

c) **Type** C_ℓ:

$$S = \{s_1, \ldots, s_\ell\}$$
$$s_i s_j \neq s_j s_i \quad \text{if } |i-j| = 1$$
$$J_0 = \{s_2, s_3, s_4, \ldots, s_\ell\}$$

$$\lambda^*(\Lambda \backslash 0) = \{\phi; \{s_1\}; \{s_1, s_2\}; \ldots; \{s_1, \ldots, s_i\}; \ldots, \{s_1, \ldots, s_\ell\}\}.$$

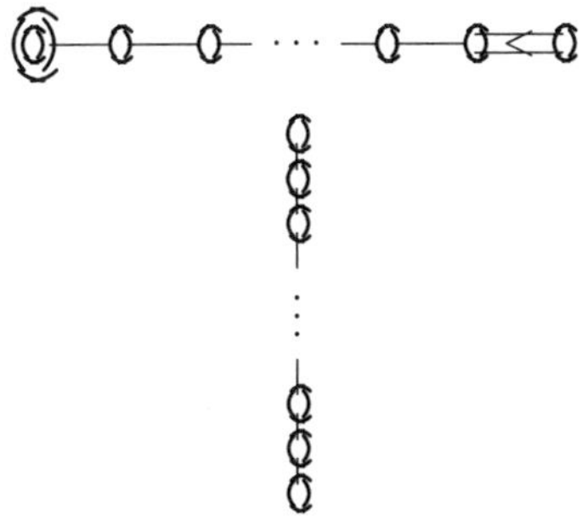

d) **Type** D_ℓ:

$$S = \{s_1, \ldots, s_{\ell-2}, s_{\ell-1}, s_\ell\}$$
$$s_i s_j \neq s_j s_i \quad \text{if } |i-j| = 1 \text{ and } i, j \leq \ell - 1 \text{ or } \{i, j\} = \{\ell - 2, \ell\}.$$
$$J_0 = \{s_1, s_3, s_4, \ldots, s_\ell\}$$

$$\begin{aligned}\lambda^*(\Lambda \backslash 0) = \{&\phi; \{s_2\}; \{s_1, s_2\}, \{s_2, s_3\}; \{s_1, s_2, s_3\}, \{s_2, s_3, s_4\}; \ldots; \\ &\{s_1, s_2, \ldots, s_{\ell-3}\}, \{s_2, s_3, \ldots, s_{\ell-2}\}; \\ &\{s_1, s_2, \ldots, s_{\ell-2}\}, \{s_2, s_3, \ldots, s_{\ell-2}, s_\ell\}, \{s_2, s_3, \ldots, s_{\ell-1}\}; \\ &\{s_1, s_2, \ldots, s_{\ell-2}, s_\ell\}, \{s_1, s_2, \ldots, s_{\ell-1}\}, \{s_2, s_3, \ldots, s_\ell\}; \\ &\{s_1, s_2, \ldots, s_\ell\}\}.\end{aligned}$$

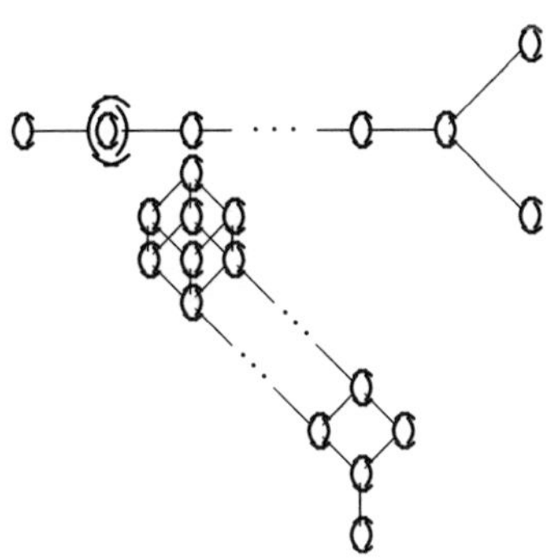

e$_6$) **Type** E_6:

$$
\begin{aligned}
&S = \{s_1, s_2, s_3, s_4, s_5, s_6\}\\
&s_i s_j \neq s_j s_i \quad \text{if } \{i,j\} \in \{\{1,2\},\{2,3\},\{3,4\},\{3,5\},\{5,6\}\}\\
&J_0 = \{s_1, s_2, s_3, s_5, s_6\}
\end{aligned}
$$

$$
\begin{aligned}
\lambda^*(\Lambda\backslash 0) = \{&\phi; \{s_4\}; \{s_3, s_4\}; \{s_2, s_3, s_4\}, \{s_3, s_4, s_5\};\\
&\{s_1, s_2, s_3, s_4\}, \{s_2, s_3, s_4, s_5\}, \{s_3, s_4, s_5, s_6\};\\
&\{s_1, s_2, s_3, s_4, s_5\}, \{s_2, s_3, s_4, s_5, s_6\}; \{s_1, s_2, s_3, s_4, s_5, s_6\}\}.
\end{aligned}
$$

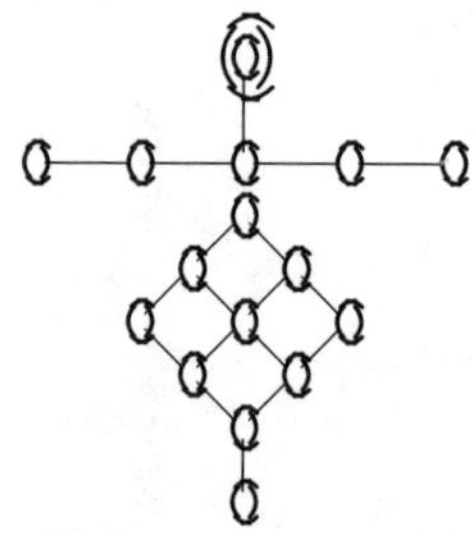

e$_7$) **Type** E_7:

$$
\begin{aligned}
&S = \{s_1, s_2, s_3, s_4, s_5, s_6, s_7\}\\
&s_i s_j \neq s_j s_i \quad \text{if } \{i,j\} \in \{\{1,2\},\{2,3\},\{3,4\},\{4,5\},\{4,6\},\{6,7\}\}\\
&J_0 = \{s_2, s_3, s_4, s_5, s_6, s_7\}
\end{aligned}
$$

$$
\begin{aligned}
\lambda^*(\Lambda\backslash 0) = \{&\phi; \{s_1\}; \{s_1, s_2\}, \{s_1, s_2, s_3\}; \{s_1, s_2, s_3, s_4\}, \{s_1, s_2, s_3, s_5\};\\
&\{s_1, s_2, s_3, s_4, s_5\}, \{s_1, s_2, s_3, s_5, s_6\};, \{s_1, s_2, s_3, s_4, s_5, s_6\},\\
&\{s_1, s_2, s_3, s_5, s_6, s_7\}; \{s_1, s_2, s_3, s_4, s_5, s_6, s_7\}\}.
\end{aligned}
$$

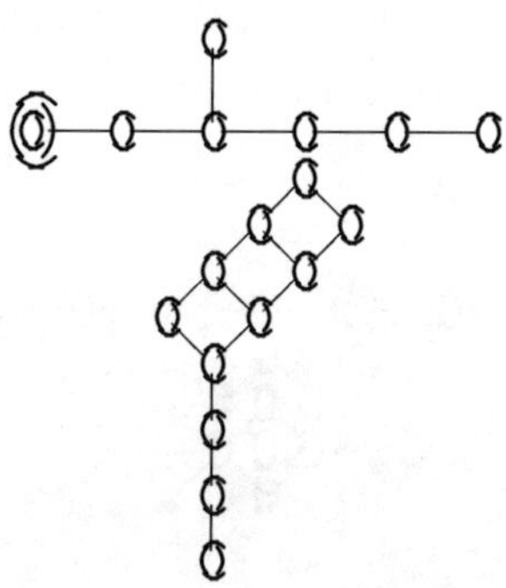

e$_8$) **Type** E_8:

$$
\begin{aligned}
&S = \{s_1, s_2, s_3, s_4, s_5, s_6, s_7, s_8\}\\
&s_i s_j \neq s_j s_i \quad \text{if } \{i,j\} \in \{\{1,2\},\{2,3\},\{3,4\},\{4,5\},\{5,6\},\{5,7\},\{7,8\}\}\\
&J_0 = \{s_2, s_3, s_4, s_5, s_6, s_7, s_8\}
\end{aligned}
$$

$$\lambda^*(\Lambda\backslash 0) = \{\phi; \{s_1\}; \{s_1, s_2\}; \{s_1, s_2, s_3\}; \{s_1, s_2, s_3\}; \{s_1, s_2, s_3, s_4\}, \{s_1, s_2, s_3, s_4, s_5\}; \{s_1, s_2, s_3, s_4, s_5, s_6\}, \{s_1, s_2, s_3, s_4, s_5, s_7\}; \{s_1, s_2, s_3, s_4, s_5, s_6, s_7\}, \{s_1, s_2, s_3, s_4, s_5, s_7, s_8\}; \{s_1, s_2, s_3, s_4, s_5, s_6, s_7, s_8\}\}.$$

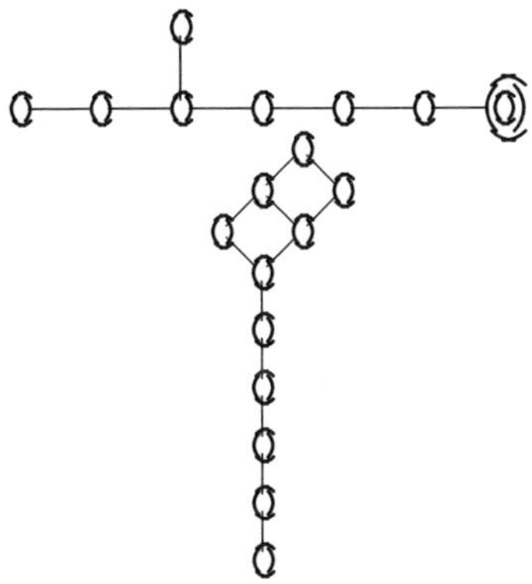

f) **Type** F_4:

$$S = \{s_1, s_2, s_3, s_4\}$$
$$s_i s_j \neq s_j s_i \quad \text{if } |i - j| = 1$$
$$J_0 - \{s_2, s_3, s_4\}$$

$$\lambda^*(\Lambda\backslash 0) = \{\phi; \{s_1\}; \{s_1, s_2\}; \{s_1, s_2, s_3\}; \{s_1, s_2, s_3, s_4\}\} .$$

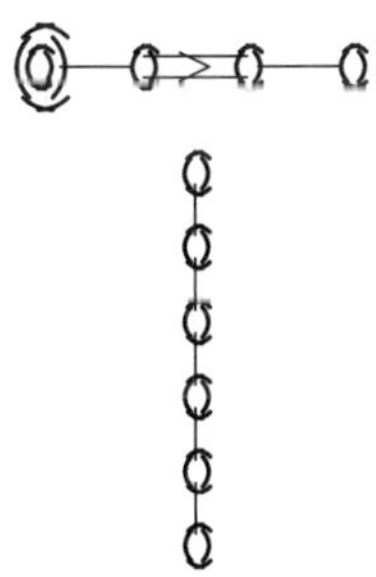

g) **Type** G_2:

$$S = \{s_1, s_2\}$$
$$s_1 s_2 \neq s_2 s_1$$
$$J_0 = \{s_1\}$$

$$\lambda^*(\Lambda\backslash 0) = \{\phi; \{s_1\}; \{s_1, s_2\}\} .$$

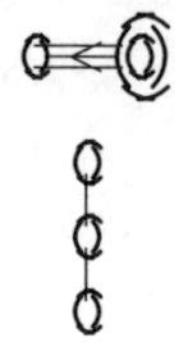

In each of these examples one may also interpret $\Lambda\backslash\{0\}$ as the lattice of centers of unipotent radicals of standard parabolic subgroups.

7.4.2 Further Examples of the Type Map

In this subsection we illustrate Theorem 7.12 with two classes of pictorial diagrams, Figure 7.1 and Figure 7.2.

In Figure 7.1 we calculate (λ, Λ) for all $\mathfrak{J}$-irreducible monoids with $J_0 = S\backslash\{s\}$. These cross section lattices correspond to $\mathfrak{J}$-irreducible monoids that arise from dominant weights μ of the form $\mu = a\omega$, where ω is a fundamental dominant weight.

In Figure 7.2 we calculate (λ, Λ) for all $\mathfrak{J}$-irreducible monoids with $J_0 = S\backslash\{s,t\}$ and $st \neq ts$. These cross section lattices correspond to $\mathfrak{J}$-irreducible monoids that arise from dominant weights μ of the form $\mu = a\omega_1 + b\omega_2$ where ω_1 and ω_2 are adjacent fundamental dominant weights. See Exercise 3 of 7.7.1 for another characterization of this class of $\mathfrak{J}$-irreducible monoids.

The structure of $\mathfrak{J}$-irreducible monoids has a peculiar, but interesting relationship with irreducible representations. The following result is originally due to S. Smith [126]. It becomes useful in the development of Putcha's abstract theory of monoids of Lie type. See Chapter 10.

Proposition 7.14. *Let $\rho : G \longrightarrow G\ell(V)$ be an irreducible representation and let $P < G$ be parabolic with $U = R_u(P)$ such that $B \subseteq P$. Let $M = \overline{K^*\rho(G)}$ be the associated $\mathfrak{J}$-irreducible monoid, with Λ, T and B as usual. Then*

a) V^U is an irreducible $P/R_u(P)$-module
b) $e \mapsto V^{R_u(P(e))}$ is a 1-1 correspondence between $\Lambda\backslash\{0\}$ and $\{V^{R_u(P)} \mid P \supseteq B\}$.

Proof. Assume first that $P = P(e)$ for some $e \in \Lambda\backslash\{0\}$. Let $W = V^U$. Since V is irreducible, we have that $W^{B_u/U} = V^{B_u}$ is one-dimensional. Hence W is an indecomposable P/U-module. Now $e \in \overline{C_G(e)}$ and so $e : W \to W$ is a $C_G(e)$-module homomorphism. Hence

$$W = e(W) \oplus ker(e)$$

as $C_G(e)$-modules. But, as already mentioned, W is indecomposable. Hence $ker(e) = 0$ and $W = e(W) = e(V)$. But $e(V)$ is irreducible over eMe by an easy calculation as in Proposition 5.1 of [95]. Thus, $e(V) = V^U$ is irreducible over P.

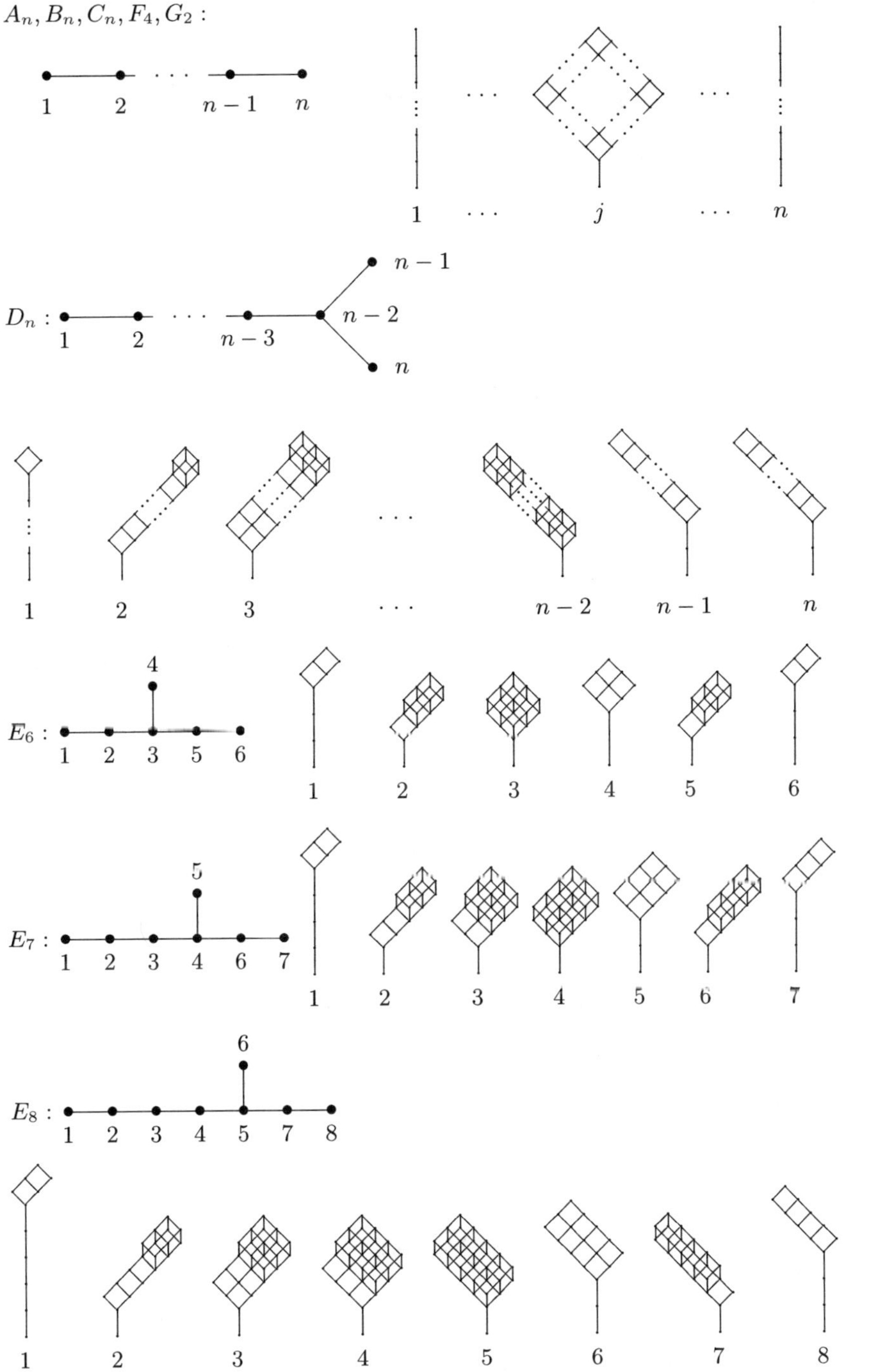

Fig. 7.1. $\Lambda \backslash \{0\}$ for $\mathcal{J}_2$-irreducible monoids. This is the case where $J_0 = S \backslash \{s\}$. Each lattice is labeled by $S \backslash J_0$.

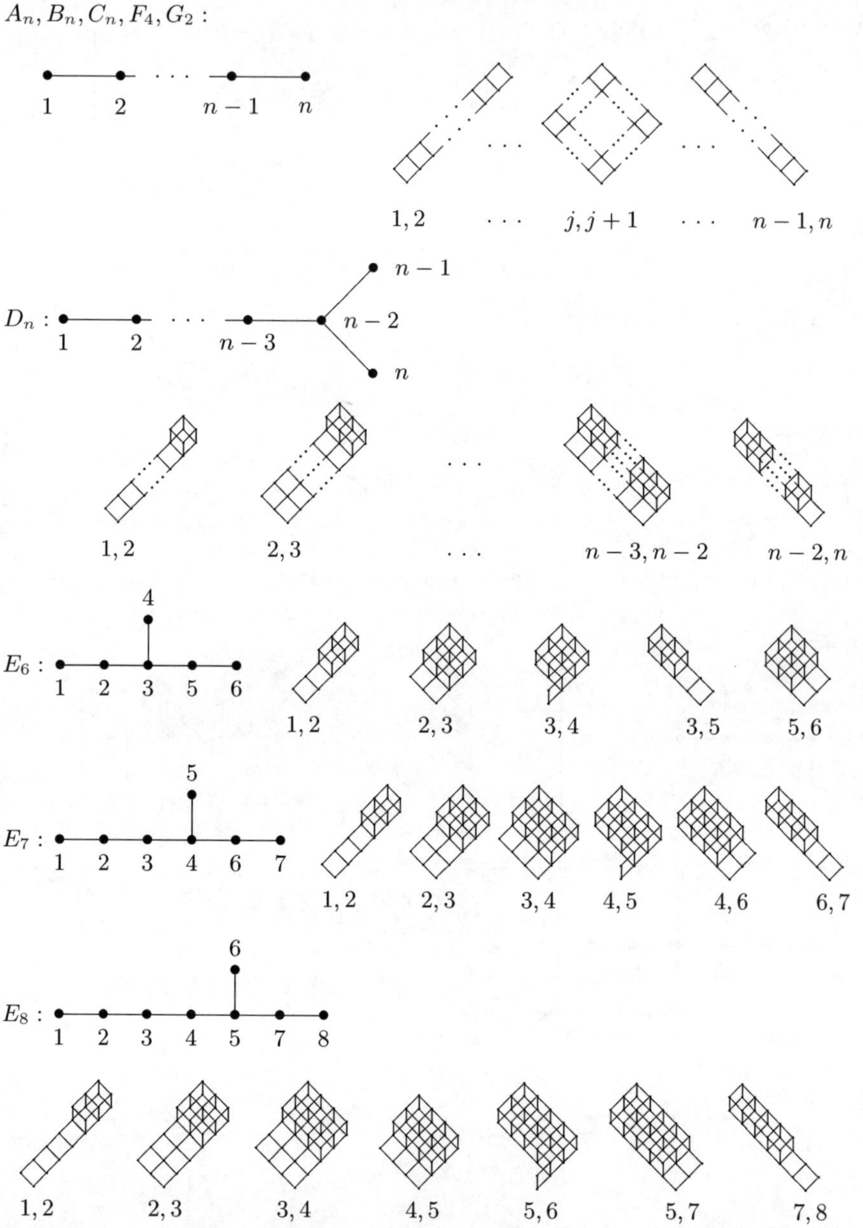

Fig. 7.2. $\Lambda\backslash\{0\}$ for $\mathcal{J}$-irreducible monoids with $J_0 = S\backslash\{s,t\}$ and $st \neq ts$. Again each lattice is labeled by $S\backslash J_0$.

Now assume that $P < G$ is any standard parabolic subgroup. We can then find $e \in \Lambda\backslash\{0\}$ such that $eC_G(e) = eC_P(e)$ and $P \subseteq P(e)$. Indeed, let $P = P_I$, and write $I = \lambda^* \sqcup \lambda_*$ where $\lambda_* \subseteq J_0$ consists of all connected components of I lying in J_0. Then $e \in \Lambda\backslash\{0\}$ is the unique idempotent with $\lambda^*(e) = \lambda^*$. Hence

$$e(V) = V^{R_u(P(e))} \supseteq V^{R_u(P)} .$$

But $R_u(P)e = \{e\}$ and so $e(V) \subseteq V^{R_u(P)}$ also. This completes the proof.

It turns out that any subspace of V, of the form V^U (where U is the unipotent radical of some standard parabolic subgroup $Q = P_I$), is already of the form $V^U = e(V)$ where $e \in \Lambda\backslash\{0\}$. In fact, $e \in \Lambda\backslash\{0\}$ is the unique minimal element of $\{e \in \Lambda\backslash\{0\} | se = es \text{ for all } s \in I\}$. For more detail, see Corollary 5.4 of [95].

7.5 2-reducible Reductive Monoids

In this section we study the orbit structure of semisimple algebraic monoids with exactly two nonzero minimal orbits. These results were first obtained in a joint paper with Putcha [98].

The case of one minimal orbit was diccussed in the previous section. The present situation is more complicated, but our results are still very precise and revealing. We associate with each 2-reducible monoid M, certain invariants (I_+, I_-) and (Δ_+, Δ_-). These invariants are not entirely independent, but should be regarded as the minimal information needed to determine the much sought after type map of M. We end the discussion with two carefully chosen examples. The first one illustrates how the Cartan matrix is used in calculating (Δ_+, Δ_-) from (I_+, I_-) and the polytope of M.

Vinberg obtained a similar description of the $G \times G$-orbits of his universal, flat deformation monoid $Env(G_0)$ of the semisimple group G_0. See § 6.3 for a summary of these results.

A reductive monoid M is **2-reducible** if $M\backslash\{0\}$ has exactly two minimal $G \times G$-orbits. Given a 2-reducible monoid M, we obtain certain invariants (I_+, I_-) and (Δ_+, Δ_-). From these, we calculate the cross section lattice Λ, and the type map of M. But (I_-, I_+) and (Δ_+, Δ_-) are not entirely independent; and it appears that the final answer depends on the "shape" of the inverse of the Cartan matrix; and not just the shape of the Dynkin diagram.

7.5.1 Reductive Monoids and Type Maps

Let M be reductive with unit group G, and let Λ be the cross-section lattice of M, relative to T and B. Then

$$\Lambda = \{e \in E(\overline{T}) \mid Be = eBe\}.$$

We let

$$\Lambda' = \Lambda \backslash \{0\},$$

so that by Theorem 4.5 c)

$$M = \bigsqcup_{e \in \Lambda} GeG.$$

Since Λ is a lattice, it has two operations, the **meet** $\wedge$, and the **join** $\vee$. The meet of $e, f \in \Lambda$ is just there product ef in M. The join of e and f is the smallest idempotent $h \in \Lambda$ with $he = e$ and $hf = f$.

In this section, we regard the type map

$$\lambda : \Lambda \to 2^{\Delta}$$

as taking values in the set of subsets of Δ, the set of simple roots. This slight change of notation should not cause difficulties.

Lemma 7.15. *Let $W(e) = W_{\lambda(e)} = \{w \in W \mid we = ew\}$, the associated parabolic subgroup of W. Let $W_*(e) = \{w \in W \mid we = ew = e\}$, and $W^*(e) = \{w \in W \mid we = ew \neq e\}$. Then*

a) $\lambda_*(e) = \bigcap_{f \leq e} \lambda(f)$ and $W_*(e) = \bigcap_{f \leq e} W(f)$;
b) $\lambda^*(e) = \bigcap_{f \geq e} \lambda(f)$ and $W^*(e) = \bigcap_{f \geq e} W(f)$.

It follows from Lemma 7.15 that

i) for $e \in \Lambda$, $\lambda(e) = \lambda^*(e) \sqcup \lambda_*(e)$;
ii) for $e, f \in \Lambda$, $\lambda(e) \cap \lambda(f) \subseteq \lambda(e \vee f) \cap \lambda(e \wedge f)$;
iii) for $e \in \Lambda$, $W(e) = W^*(e) \times W_*(e)$;
iv) if $e \geq f$ then $\lambda_*(e) \subseteq \lambda_*(f)$ and $\lambda^*(f) \subseteq \lambda^*(e)$. Furthermore, λ^* restricted to eMe is the λ^* of eMe, and λ_* restricted to M_e is the λ_* of M_e.

Here, eMe is the reductive monoid with unit group $eC_G(e)$, and $M_e = \overline{G_e}$, where $G_e = \{g \in G \mid ge = eg = e\}^0$; M_e is also a reductive monoid.

Definition 7.16. *Let M, Λ and λ be as above. Let $\Lambda_1 \subseteq \Lambda$ be the subset of nonzero minimal elements.*

a) The core *C of Λ is*

$$C = \{e \in \Lambda \mid e = e_1 \vee \cdots \vee e_k, \textit{ for some } e_i \in \Lambda_1\}.$$

b) Define $\theta : \Lambda' \to C$ by

$$\theta(e) = \vee\{f \in \Lambda_1 \mid f \leq e\}$$

so that, if $e_1 \leq e_2$, then $\theta(e_1) \leq \theta(e_2)$.

c) Write $\Lambda' = \bigsqcup_{h \in C} \Lambda_h$, where $\Lambda_h = \theta^{-1}(h)$

Recall that a reductive monoid M with zero is *semisimple* if $dim(Z(G)) = 1$. For any semisimple monoid M, there is a special relationship between Λ' and $S \subseteq X(T)$. If $\alpha \in \Delta$ then by Theorem 10.20 of [82] there exists $e_\alpha \in \Lambda'$ such that $P(e_\alpha) = P_{\Delta\setminus\{\alpha\}}$. Morover, e_α is unique. See Lemma 7.22 below.

Definition 7.17. *Let $C \subseteq \Lambda'$ be the core.*

a) Define $\pi : \Delta \to C$ by $\pi(\alpha) = \theta(e_\alpha)$.
b) Write $\Delta = \sqcup_{h \in C} \Delta_h$, where $\Delta_h = \pi^{-1}(h)$.

Proposition 7.18. *a) If $e \in \Lambda_h$ then*

$$\lambda_*(e) = \{\alpha \in \lambda_*(h) \mid s_\alpha s_\beta = s_\beta s_\alpha \textit{ for all } \beta \in \lambda^*(e)\}.$$

b) If $e \in \Lambda_h$ and $f \in \Lambda_k$ then

$$e \leq f \textit{ if and only if } h \leq k \textit{ and } \lambda^*(e) \subseteq \lambda^*(f).$$

Proof. Consider a). Since $e \geq h$, $\lambda_*(e) \subseteq \lambda_*(h)$. Let $\alpha \in \lambda^*$. Then since $W(e) = W^*(e) \times W_*(e)$, $s_\alpha s_\beta = s_\beta s_\alpha$ for all $\beta \in \lambda^*(e)$.

So it remains to prove the reverse inclusion. Now $E(e\overline{T})$ is the face lattice of a polytope (see Section 4). Therefore e is the join of the nonzero minimal idempotents of $E(e\overline{T})$. Hence

$$e = \vee\{xe'x^{-1} \mid e' \in \Lambda_1,\ e \geq e',\ x \in W^*(e)\}.$$

Let $\alpha \in \lambda_*(h)$ be such that $s_\alpha s_\beta = s_\beta s_\alpha$ for all $\beta \in \lambda^*(e)$. Then $s_\alpha x = x s_\alpha$ for all $x \in W^*(e)$. Let $e' \in \Lambda_1$ be such that $e \geq e'$. Since $e \in \Lambda_h, h \geq e'$. Let $x \in W^*(e)$. However, $\alpha \in \lambda^*(h)$. Hence

$$s_\alpha x e' x^{-1} = x s_\alpha e' x^{-1} = x e' x^{-1} = x e' s_\alpha x^{-1} = x e' x^{-1} s_\alpha.$$

By the above join formula for e, and Proposition 7.5 of [82], $s_\alpha x = x s_\alpha$. Thus, $s_\alpha \in W(e)$. Now s_α commutes with all the nonzero minimal idempotents in $E(e\overline{T})$, and thus, es_α has the same property. Thus, es_α commutes with all idempotents of of $e\overline{T}$. Since $eW(e)$ acts faithfully on $E(e\overline{T})$, it follows from Chapter 10 of [82] that $es_\alpha = e$. Hence $s_\alpha \in W_*(e)$ and $\alpha \in \lambda_*(e)$.

For b), let $h \leq k$ and $\lambda^*(e) \subseteq \lambda^*(f)$. Let $e' \in \Lambda_1$ be such that $e \geq e'$. Then $e' \leq h \leq k \leq f$. Let $x \in W^*(e) \subseteq W^*(f)$. Then,

$$fxe'x^{-1} = xfe'x^{-1} = xe'x^{-1}.$$

Hence $xe'x^{-1} \leq f$. Therefore by the above join formula for e, $e \leq f$. The converse is clear.

7.5.2 The Type Map of a 2-reducible Monoid

Let M be a 2-reducible, semisimple monoid. Our terminology is well chosen because of the following proposition.

Proposition 7.19. *Let M be a semisimple monoid. The following are equivalent.*

a) M is 2-reducible;
b) i) there is a rational representation $\rho : M \to End(V \oplus W)$ so that ρ is finite as a morphism, and V and W are irreducible M-summands;
ii) M is not $\mathfrak{Z}$-irreducible.

Proof. If M is 2-reducible, let $\Lambda_1 = \{e, f\}$. There exist irreducible representations $\rho_1 : M \to End(V)$ and $\rho_2 : M \to End(W)$ such that $\rho_1(e) \neq 0$ and $\rho_2(f) \neq 0$. It is easy to check that $\rho = \rho_1 \oplus \rho_2$ does the job. Conversely, if the conditions of b) are satisfied, let $\Lambda_1 = \{e_1, \ldots, e_r\}$, where $r \geq 2$. We can assume that $e_1(V) \neq 0$. But then $e_1(V)$ generates V as an M-module, and so $e_2(V) = 0$ since $e_1 G e_2 = 0$. Thus $e_2(W) \neq 0$. But now for any $i > 2$, $e_i(V \oplus W) = 0$. Thus $r = 2$.

In this section, we determine Λ and $\lambda : \Lambda \to 2^{\Delta}$ in terms of certain invariants (I_+, I_-) and (Δ_+, Δ_-).

Write

$$\Lambda_1 = \{e_+, e_-\}.$$

Then

$$C = \{e_+, e_-, e_0\},$$

where $e_0 = e_+ \vee e_-$. Let

$$I_+ = \lambda_*(e_+),\ I_- = \lambda_*(e_-) \text{ and } I_0 = \lambda_*(e_0).$$

Then

$$I_0 = I_+ \cap I_-.$$

By 7.16 c),

$$\Lambda' = \Lambda_+ \sqcup \Lambda_- \sqcup \Lambda_0$$

and by 7.17 b)

$$\Delta = \Delta_+ \sqcup \Delta_- \sqcup \Delta_0$$

where $\Delta_+ = \pi^{-1}(e_+), \Delta_- = \pi^{-1}(e_-)$, and $\Delta_0 = \pi^{-1}(e_0)$. Hence

i) $\alpha \in \Delta_+$ if $e_\alpha \geq e_+$ and $e_\alpha \ngeq e_-$;
ii) $\alpha \in \Delta_-$ if $e_\alpha \geq e_-$ and $e_\alpha \ngeq e_+$;
iii) $\alpha \in \Delta_0$ if $e_\alpha \geq e_+$ and $e_\alpha \geq e_-$.

See the paragraph preceding Definition 7.17 the definition of e_α. By Proposition 7.18, our problem is reduced to determining $\lambda^*(\Lambda_+), \lambda^*(\Lambda_-)$ and $\lambda^*(\Lambda_0)$.

Remark 7.20. If M is 2-reducible but not semisimple, then $dim(Z(G)) = 2$. One can then show that, in this situation,

a) $\lambda^*(\Lambda_+) = \{X \subseteq \Delta |$ no component of X is in $I_+\}$
b) $\lambda^*(\Lambda_-) = \{X \subseteq \Delta |$ no component of X is in $I_-\}$
c) $\lambda^*(\Lambda_0) = \{X \subseteq \Delta |$ no component of X is in $I_0\}$.

In this case, M is a special case of the *multilined closure* with $n = 2$. Here n is the number of minimal $G \times G$ orbits of $M \backslash \{0\}$. The multilined closure is an appealing situation where the lattice of orbits and the type map can be written down directly in terms of the types of the minimal orbits. This has been described in generality in Chapter 6. See also Remark 7.21 below for the case $n = 1$. In any case, the semisimple case is more complicated. It is also more interesting.

Remark 7.21. We shall freely use the results from § 7.3 about $\mathfrak{J}$-irreducible monoids in the proof of Theorem 7.23 below. Notice that Proposition 7.18 includes Theorem 7.12 as a special case.

We now return to the 2-reducible case.

Lemma 7.22. *$\Delta_+ \neq \phi$ and $\Delta_- \neq \phi$.*

Proof. Choose a maximal $e \in \Lambda_+$. Then e is covered by some $f \in \Lambda_0$. Furthermore, f is unique, since if e is also covered by $h \in \Lambda_0$, and $f \neq h$, then $e = fh \geq e_0$, a contradiction. Thus, both fMf and M_f are $\mathfrak{J}$-irreducible, and hence semisimple. Hence $\lambda(e) = \Delta\backslash\{\alpha\}$ for some $\alpha \in \Delta$. This $e \in \Lambda$ is actually unique with $\lambda(e) = \Delta\backslash\{\alpha\}$. (The connnected center Z of $C_G(e)$ is two dimensional. So $\overline{Z}$ has exactly four idempotents $\{e, f, 0, 1\}$. $P(f)$ is the opposite parabolic of $P(e)$. But then $B \not\subseteq P(f)$, so that $f \notin \Lambda$.) In any case, $\alpha \in \Delta_+$. Similarly, $\Delta_- \neq \phi$.

As we already mentioned, we want to determine $\lambda : \Lambda \to 2^\Delta$ in terms of I_+, I_-, Δ_+ and Δ_-. By Proposition 2.5, it suffices to determine the sets $\lambda^*(\Lambda_+), \lambda^*(\Lambda_-)$ and $\lambda^*(\Lambda_0)$. Let

$\mathcal{A}_+ = \{X \subseteq \Delta \mid$ no component of X is contained in $I_+,\ \Delta_+ \not\subseteq X\}$
$\mathcal{A}_- = \{X \subseteq \Delta \mid$ no component of X is contained in $I_-,\ \Delta_- \not\subseteq X\}$
$\mathcal{A}_0 = \{X \subseteq \Delta \mid$ no component of X is contained in I_0, and either $\Delta_+ \not\subseteq X$ and $\Delta_- \not\subseteq X$ or else $\Delta_+ \cup \Delta_- \subseteq X\}$.

Theorem 7.23.
a) $\lambda^*(\Lambda_+) = \mathcal{A}_+$;
b) $\lambda^*(\Lambda_-) = \mathcal{A}_-$;
c) $\lambda^*(\Lambda_0) = \mathcal{A}_0$.
In all cases, λ^* is injective.

Proof. Suppose first that $X \in \mathcal{A}_+$. Then $\alpha \notin X$ for some $\alpha \notin \Delta_+$. By Theorem 10.20 of [82], there exists $e \in \Lambda_+$ such that $\lambda(e) = \Delta\backslash\{\alpha\}$. Hence $X \subseteq \lambda(e)$. By Proposition 7.18 a), $X \subseteq \lambda^*(e)$. Now eMe is a $\mathcal{J}$-irreducible monoid of type $I_+ \cap \lambda^*(e)$. Since no component of X is contained in I_+, there exists $f < e$ such that $\lambda^*(f) = X$. Clearly, $f \in \Lambda_+$.

Conversely, let $f \in \Lambda_+$. Let $e \in \Lambda_+$ be maximal such that $f \le e$. By the proof of Lemma 7.22, $|\lambda(e)| = |\Delta| - 1$. Hence $\lambda(e) = \Delta\backslash\{\alpha\}$ for some $\alpha \in \Delta_+$. Also, $\lambda_*(f) \subseteq \lambda_*(e) \subseteq \Delta\backslash\{\alpha\}$ by iv) following Definition 7.16. Hence $\lambda^*(f) \in \mathcal{A}_+$, by Proposition 7.18 a).

Similarly, $\lambda^*(\Lambda_-) = \mathcal{A}_-$.

To prove c), we proceed by induction on $dim(M)$. Let $f \in \Lambda_0$. Then $f \le e$ for some maximal $e \neq 1$. So eMe is a 2-reducible monoid. First, suppose that eMe is not semisimple. Then by Remark 7.20, $\lambda(e) = \lambda^*(e) = \Delta\backslash\{\alpha_1, \alpha_2\}$. By Proposition 7.18, $\alpha_1, \alpha_2 \notin \Delta_0$. Suppose that $\alpha_1, \alpha_2 \in S_+$. Then there exist $e_1, e_2 \in \Lambda_+$ such that $\lambda(e_1) = \Delta\backslash\{\alpha_1\}$ and $\lambda(e_2) = \Delta\backslash\{\alpha_2\}$. By Remark 7.20, there exists $h \in \Lambda_+$ such that e covers h and $\lambda(h) \subseteq \Delta\backslash\{\alpha_1, \alpha_2\}$. By Proposition 7.18 b), $h < e_1$ and $f < e_2$. But $\{1, e_2, h\}$ is a maximal chain in $E(\overline{T}_h)$. So $dim(T_h) = 2$, while $\{1, e_1, e_2, e, h\} \subseteq E(\overline{T}_h)$. This is a contradiction since $|E(\overline{T}_h)| = 4$ for such D-monoids. Similarly, $\alpha_1, \alpha_2 \in \Delta_-$ leads to a contradiction. So assume that $\alpha_1 \in \Delta_+$ and $\alpha_2 \in \Delta_-$. Then by Proposition 7.18 b), $\lambda^*(f) \subseteq \lambda^*(e) = \Delta\backslash\{\alpha_1, \alpha_2\}$. Hence $\lambda^*(f) \in \mathcal{A}_0$.

Next assume that eMe is semisimple. Then $\lambda(e) = \lambda^*(e) = \Delta\backslash\{\beta\}$ for some $\beta \in \Delta_0$. Correspondingly, in eMe, let

$$\Delta\backslash\{\beta\} = \Delta'_+ \sqcup \Delta'_0 \sqcup \Delta'_-.$$

Let λ_1 denote λ in eMe. We claim that $\Delta_+ = \Delta'_+$. Let $\alpha \in \Delta_+$. Since eMe is a semisimple monoid, there exists $e_1 < e$ such that $\lambda_1(e_1) = \Delta\backslash\{\alpha, \beta\}$. If $\lambda(e_1) = \Delta\backslash\{\alpha\}$, then $e_1 \in \Lambda_+$, and hence $\alpha \in \Delta'_+$. So assume that $\lambda(e_1) = \Delta\backslash\{\alpha, \beta\}$. Now $\lambda(e_2) = \Delta\backslash\{\alpha\}$ for some $e_2 \in \Lambda_+$. However,

$$\beta \notin \lambda^*(e_2) \implies \lambda^*(e_2) \subseteq \Delta\backslash\{\beta\} \implies e_2 \le e.$$

But

$$e_2 \le e \implies \lambda_1(e_2) = \Delta\backslash\{\alpha, \beta\} \implies e_1 = e_2 \implies \alpha \in \Delta'_+$$

Therefore let $\beta \in \lambda^*(e_2)$. Since $e_2 \in \Lambda_+$, e_2Me_2 is $\mathcal{J}$-irreducible, and hence semisimple. Let λ_2 denote λ for e_2Me_2. There exists $e_3 < e_2$ such that $\lambda_2(e_3) = \lambda_2(e_2)\backslash\{\beta\}$. hence

$$\Delta\backslash\{\alpha, \beta\} = \lambda_*(e_2) \cup (\lambda^*(e_2)\backslash\{\beta\}) \subseteq \lambda_2(e_3) \cup \lambda_*(e_2) \subseteq \lambda(e_3).$$

If $\lambda(e_3) = \Delta\backslash\{\beta\}$, then $e_3 = e \in \Lambda_0$, a contradiction. Hence $\lambda(e_3) = \Delta\backslash\{\alpha, \beta\}$. By Proposition 7.18 b), $e_3 < e$ and so $\alpha \in \Delta'_+$. Thus, $\Delta_+ \subseteq \Delta'_+$. Similarly, $\Delta_- \subseteq \Delta'_-$.

Suppose that $\alpha \in \Delta'_+$, $\alpha \notin \Delta_+$. Then $\alpha \notin \Delta_-$ since $\Delta_- \subseteq \Delta'_-$. Hence $\alpha \in \Delta_0$. There exists $e_1 \in \Lambda_+$ with $e_1 < e$ such that $\lambda_1(e_1) = \Delta\backslash\{\alpha, \beta\}$. Since

$\alpha \notin \Delta_+$, $\lambda(e_1) \neq \Delta\backslash\{\alpha\}$. Hence $\lambda(e_1) = \Delta\backslash\{\alpha,\beta\}$. Now $\lambda(e_2) = \Delta\backslash\{\alpha\}$ for some $e_2 \in \Lambda_0$, since $\alpha \in \Delta_0$. By Proposition 7.18 b), $e_1 < e_2$. Hence $e_1 < ee_2$. By ii) of § 7.5.1

$$\Delta\backslash\{\alpha,\beta\} = (\Delta\backslash\{\alpha\}) \cap (\Delta\backslash\{\beta\}) \subseteq \lambda(ee_2).$$

By Proposition 7.18 b), $\lambda(ee_2) \neq \Delta\backslash\{\alpha\}$ or $\Delta\backslash\{\beta\}$. Hence $\lambda(ee_2) = \Delta\backslash\{\alpha,\beta\}$. Hence $\lambda_1(e_1) = \lambda_1(ee_2) = \Delta\backslash\{\alpha,\beta\}$. Since eMe is semisimple, $e_1 = ee_2 \in \Lambda_0$, a contradiction. Hence $\Delta'_+ \subseteq \Delta_+$ and so $\Delta'_+ = \Delta_+$. Similarly, $\Delta'_- = \Delta_-$. By the induction hypothesis, $\lambda^*(f) \in \mathcal{A}_0$. Thus, $\lambda^*(\Lambda_0) \subseteq \mathcal{A}_0$.

Conversely, let $X \in \mathcal{A}_0$. Suppose first that $\Delta_+ \cup \Delta_- \subseteq X, X \neq \Delta$. Then $X \subseteq \Delta\backslash\{\beta\}$ for some $\beta \in \Delta_0$. There exists $f \in \Lambda_0$ such that $\lambda(f) = \Delta\backslash\{\beta\}$. If fMf is semisimple, then $\Delta'_+ = \Delta_+$ and $\Delta'_- = \Delta_-$ as above; and by the induction hypothesis, $\lambda^*(f') = X$ for some $f' \in \Lambda_0, f' \leq f$. If fMf is not semisimple, then the same is true by Remark 7.21.

Suppose next that $\Delta_+ \nsubseteq X$ and $\Delta_- \nsubseteq X$. Let $\alpha \in \Delta_+, \beta \in \Delta_-$ be such that $X \subseteq \Delta\backslash\{\alpha,\beta\}$. We first show that there exists $f \in \Lambda_0$ such that $\lambda(f) = \Delta\backslash\{\alpha,\beta\}$. Now there exists $e \in \Lambda_+$ such that $\lambda(e) = \Delta\backslash\{\alpha\}$. Then M_e and eMe are both semisimple. Suppose that $\beta \in \lambda_*(e)$. Then there exists $f > e$ such that $\lambda_1(f) = \lambda^*(e)\backslash\{\beta\}$, where λ_1 is λ for M_e. So in M (using iv) following Lemma 7.15),

$$\Delta\backslash\{\alpha,\beta\} = (\lambda_*(e)\backslash\{\beta\}) \cup \lambda^*(e) \subseteq \lambda(f).$$

Since $f > e \geq e_+$, $f \notin \Lambda_-$. Hence, $\lambda(f) \neq \Delta\backslash\{\beta\}$ and so $\lambda(f) = \Delta\backslash\{\alpha,\beta\}$. So e is central in fMf, and thus fMf is not $\mathfrak{J}$-irreducible. Hence, $f \notin \Lambda_+$. Thus $f \in \Lambda_0$.

Assume next that $\beta \in \lambda^*(e)$. Then there exists $e_1 < e$ such that $\lambda_1(e_1) = \lambda_*(e)\backslash\{\beta\}$, where λ_1 is λ for eMe. So by iv) just following Lemma 7.15,

$$\Delta\backslash\{\alpha,\beta\} = (\lambda^*(e)\backslash\{\beta\}) \cup \lambda_*(e) \subseteq \lambda(e_1).$$

Since $e_1 < e, e_1 \notin \Lambda_-$. Hence $\lambda(e_1) \neq \Delta\backslash\{\beta\}$ and so $\lambda(e_1) = \Delta\backslash\{\alpha,\beta\}$. Hence e is central in M_{e_1}. Thus M_{e_1} has at least four central idempotents. So let f be a central idempotent of M_{e_1} such that $f \notin \{1, e, e_1\}$. Then $\Delta\backslash\{\alpha,\beta\} \subseteq \lambda(f)$ by iv) again. Since $f > e_1$, $f \notin \Lambda_-$ and so $\lambda(f) \neq \Delta\backslash\{\beta\}$. Since $f \neq e$, $\lambda(f) \neq \Delta\backslash\{\alpha\}$. Thus $\lambda(f) = \Delta\backslash\{\alpha,\beta\}$. If $f \in \Lambda_+$, then fMf is $\mathfrak{J}$-irreducible, and e_1 is a central idempotent: a contradiction. Hence $f \in \Lambda_0$.

There exists $f \in \Lambda_0$ such that $\lambda(f) = \Delta\backslash\{\alpha,\beta\}$. Hence either M_f is not semisimple, or fMf is not semisimple. Suppose that M_f is not semisimple. There exists $f' > f$, $f \neq 1$, such that f' is central in M_f. By iv) just after Lemma 7.15, $\Delta\backslash\{\alpha,\beta\} \subseteq \lambda(f')$. Since $f' \in \Lambda_0$, $\alpha \in \Delta_+$, $\beta \in \Delta_-$, $\lambda(f') \neq \Delta\backslash\{\alpha\}$ and $\lambda(f') \neq \Delta\backslash\{\beta\}$. Hence $\lambda(f') = \Delta\backslash\{\alpha,\beta\}$. Thus by iv) again, and Proposition 7.18 b), $f = f'$: a contradiction. Consequently M_f is semisimple. But then fMf is not semisimple. Since $X \subseteq \Delta\backslash\{\alpha,\beta\}$, $X \subseteq \lambda_*(f)$. By Remark 7.21, $\lambda^*(f') = X$ for some $f' \in \Lambda_0$, $f' \leq f$. Thus $\mathcal{A} \subseteq \lambda^*(\Lambda_0)$. This concludes the proof.

Corollary 7.24. *The partial order on Λ is determined as follows. Let $e, f \in \Lambda$. Then the following are equivalent:*

a) $e \leq f$;
b) *i)* $\lambda^*(e) \subseteq \lambda^*(f)$, *and*
ii) $e, f \in \Lambda_+$; $e, f \in \Lambda_-$; $e, f \in \Lambda_0$; $e \in \Lambda_+, f \in \Lambda_0$; *or* $e \in \Lambda_-, f \in \Lambda_0$.

Proof. This is straightforward using Proposition 7.18 and Theorem 7.23.

7.5.3 Calculating the Type Map Geometrically

In the previous section we found the exact description of the type map

$$\lambda : \Lambda \to 2^{\Delta}$$

of a 2-reducible monoid by first identifying the necessary combinatorial invariants (I_+, I_-) and $(\Delta_+, \Delta_-, \Delta_0)$. In this section we determine some geometric refinements of that situation by calculating the decomposition

$$\Delta = \Delta_+ \sqcup \Delta_- \sqcup \Delta_0$$

in terms of the coordinates of $\Lambda_1 = \{e_+, e_-\}$, thought of as vertices of the polytope $\mathcal{P}$ of M. The problem here is to determine which decompositions of Δ are possible for a 2-reducible monoid M of type (I_+, I_-). This is no longer a purely combinatorial problem.

Let M be a 2-reducible, semisimple monoid, and let T, $\overline{T}$, Λ, etc. have the usual meanings. As above, let $\mathcal{P}$ be the polytope of M. By Theorem 7.5, we have a canonical bijection

$$\iota : \Lambda_1 \to \{x, y\}.$$

We write $\iota(e_+) = x$ and $\iota(e_-) = y$ where $\{x, y\}$ is the set of vertices of $\mathcal{P}$ that are contained in $X(\overline{T})^+ \otimes \mathbb{Q}^+$. Let $Bd(\mathcal{P})$ be the boundry of $\mathcal{P}$. For $\alpha \in \Delta$ let

i) $H_\alpha = Span_{\mathbb{Q}}(\Delta \backslash \{\alpha\})$
ii) $H_\alpha^+ = Cone_{\mathbb{Q}^+}(\Delta \backslash \{\alpha\})$.

For $\alpha \in \Delta$, let $\omega_\alpha \in X(\overline{T})^+ \otimes \mathbb{Q}^+$ be the fundamental dominant weight that is orthogonal to H_α.

Lemma 7.25. *For any $\alpha \in \Delta$ there is a unique $z_\alpha \in \mathbb{Q}^+\omega_\alpha$ such that*

$$(z_\alpha + H_\alpha) \cap \mathcal{P} = (z_\alpha + H_\alpha) \cap Bd(\mathcal{P}) \neq \phi.$$

Furthermore,

i) $z_\alpha \in Bd(\mathcal{P})$
ii) $(z_\alpha + H_\alpha) \cap \mathcal{P}$ *is the face F of $\mathcal{P}$ corresponding to e_α.*

Proof. Let $e = e_\alpha$ be the unique idempotent such that $\lambda(e_\alpha) = \Delta\backslash\{\alpha\}$ (see Lemma 7.22). Let $F \in \mathcal{F}$ be the face of $\mathcal{P}$ corresponding to $e \in \Lambda$. Then

$$\mathbb{Q}^+\omega_\alpha \subseteq \mu_e,$$

and thus $\mathbb{Q}^+\omega_\alpha \cap F = \{z_\alpha\}$ (since F is a subset of $z_\alpha + H_\alpha$, it must be orthogonal to $\mathbb{Q}\omega_\alpha$). Clearly, $F \subseteq Bd(\mathcal{P})$.

Let $I = \Delta\backslash\{\alpha\}$. Then F is W_I-invariant. Thus $F - z_\alpha$ is also W_I-invariant. But $\mathbb{Q}\omega_\alpha \cap (F - z_\alpha) = \{0\}$, and so $(F - z_\alpha)^{W_I} = \{0\}$. Thus $F - z_\alpha \subseteq H_\alpha$. Hence $F \subseteq H_\alpha + z_\alpha$.

The author would like to thank Hugh Thomas for the proof of the following Lemma.

Lemma 7.26. *The following are equivalent:*

a) $x \in z_\alpha + H_\alpha$
b) $x \in z_\alpha + H_\alpha^+$
c) $e_+ \leq e_\alpha$.

The corresponding result holds with x replaced by y and e_+ replaced by e_-.

Proof. For $\alpha \in \Delta$, let $C_1 = Cone(\{\omega_\alpha\})$ and $C_2 = Cone((\Delta\backslash\{\alpha\}) \cup \{\omega_\alpha\})$. We claim that $C_1 \subseteq C_2$. It suffices to show that $\omega_\beta \in C_2$ for any $\beta \in \Delta\backslash\{\alpha\}$. Now

$$X(T) \otimes \mathbb{Q} = H_\alpha \oplus \mathbb{Q}\omega_\alpha,$$

an orthogonal decomposition. So let

$$\omega_\beta = x + c\omega_\alpha.$$

It suffice to show that

i) $c \geq 0$, and
ii) $x \in Cone(\Delta\backslash\{\alpha\})$.

To get i), we use the inner product. Since $\omega_\beta = x + c\omega_\alpha$, we obtain

$$< \omega_\beta, \omega_\alpha > = < x, \omega_\alpha > + c < \omega_\alpha, \omega_\alpha > .$$

But $< x, \omega_\alpha > = 0$, so that

$$c = < \omega_\beta, \omega_\alpha > / < \omega_\alpha, \omega_\alpha >,$$

and it is well known that this is non-negative.

To get ii), first notice that

$$< \beta, x > = < \beta, \omega_\beta - c\omega_\alpha > = < \beta, \omega_\beta > = 1.$$

But if $\gamma \neq \beta, \alpha$, we obtain

$$< \gamma, x >=< \gamma, \omega_\beta - c\omega_\alpha >= 0.$$

So x is the dual of β in the root system $(H_\alpha, \Delta\backslash\{\alpha\})$. But it is well known that, for any root system, the cone generated by the fundamental weights is contained in the cone generated by the positive roots, since the inverse of the Cartan matrix has positive entries. This proves the claim.

Now let $\mathcal{C} = Cone(\{\omega_\alpha | \alpha \in \Delta\})$. We claim now that

$$(z_\alpha + H_\alpha) \cap \mathcal{C}) = (z_\alpha + H_\alpha^+) \cap \mathcal{C}.$$

From our first claim,

$$\mathcal{C} = \bigcup_{r \geq 0} (rz_\alpha + H_\alpha^+) \cap \mathcal{C} = \mathcal{C} = \bigcup_{r \geq 0} (rz_\alpha + H_\alpha) \cap \mathcal{C}.$$

But $(rz_\alpha + H_\alpha) \cap (sz_\alpha + H_\alpha) = \phi$ if $r \neq s$. Hence

$$(z_\alpha + H_\alpha) \cap \mathcal{C}) \subseteq (z_\alpha + H_\alpha^+) \cap \mathcal{C},$$

and this establishes the second claim.

Now assume that $x \in z_\alpha + H_\alpha$. Then since $x \in \mathcal{P} \cap \mathcal{C}$, we get from the claim that $x \in z_\alpha + H_\alpha^+$. So clearly, a) and b) are equivalent. Also a) and c) are equivalent since, from Lemma 7.25, $(z_\alpha + H_\alpha) \cap \mathcal{P}$ is the face of $\mathcal{P}$ corresponding to $e_\alpha \in \Lambda$; while $x \in \mathcal{P}$ is the vertex of $\mathcal{P}$ corresponding to e_+. This completes the proof.

Corollary 7.27. *For each $\alpha \in \Delta$, either $x \in z_\alpha + H_\alpha$, or else $y \in z_\alpha + H_\alpha$.*

Proof. $\{e_\alpha, e_+, e_-\} \subseteq \Lambda'$, *while* $\Lambda_1 = \{e_+, e_-\}$. *Thus* $e_\alpha \geq e_+$ *or else* $e_\alpha \geq e_-$.

Theorem 7.28. *Write $x - y = \sum_{\alpha \in \Delta} r_\alpha \alpha$, where $r_\alpha \in \mathbb{Q}$.*

a) The following are equivalent:
- *i)* $r_\alpha > 0$
- *ii)* $e_\alpha \in \Lambda_+$
- *iii)* $x \in z_\alpha + H_\alpha^+$, $y \notin z_\alpha + H_\alpha^+$.

b) The following are equivalent:
- *i)* $r_\alpha < 0$
- *ii)* $e_\alpha \in \Lambda_-$
- *iii)* $y \in z_\alpha + H_\alpha^+$, $x \notin z_\alpha + H_\alpha^+$.

c) The following are equivalent:
- *i)* $r_\alpha = 0$
- *ii)* $e_\alpha \in \Lambda_0$
- *iii)* $x \in z_\alpha + H_\alpha^+$, $y \in z_\alpha + H_\alpha^+$.

Proof. In each case, it suffices to show that i) and ii) are equivalent since, by Lemma 7.26, ii) and iii) are equivalent. By Corollary 7.27, exactly one of a) iii), b) iii) or c) iii) occurs.

In case a), $x \in z_\alpha + H_\alpha^+$ and $y \notin z_\alpha + H_\alpha^+$. Then

$$x = z_\alpha + \sum_{\beta \neq \alpha} a_\beta \beta$$

and

$$y = z_\alpha + \sum_{\beta \in \Delta} b_\beta \beta.$$

But $b_\alpha < 0$, since y lies in the bounded part of $\mathcal{C} \backslash (z_\alpha + H_\alpha)$, and thus "below" the hyperplane $z_\alpha + H_\alpha$. Hence

$$x - y = \sum_{\beta \neq \alpha} a_\beta \beta - \sum_{\beta \in \Delta} b_\beta \beta = \sum_{\beta \neq \alpha} (a_\beta - b_\beta)\beta \ - \ b_\alpha \alpha.$$

Hence $r_\alpha = -b_\alpha > 0$ here. Case b) is similar to case a).

In case c) we can write

$$x = z_\alpha + \sum_{\beta \neq \alpha} a_\beta \beta$$

and

$$x = z_\alpha + \sum_{\beta \neq \alpha} b_\beta \beta.$$

Hence

$$x - y = \sum_{\beta \neq \alpha} (a_\beta - b_\beta)\beta$$

and thus, $r_\alpha = 0$ in this case.

7.5.4 Monoids with $I_+ = \Delta \backslash \{\alpha\}$ and $I_- = \Delta \backslash \{\beta\}$

In this section we exhibit some explicit calculations of the type maps of 2 reducible monoids. We restrict our attention to certain monoids with group $G = Gl_{n+1}(K)$. The general problem here is to determine all possible $(+, -, 0)$-decompositions of Δ that can actually occur for the given I_+ and I_-. We do not yet have a general solution to this intriguing problem. However, our calculations indicate that it has something to do with linear programming problems involving the inverse of the Cartan matrix.

So let $G = Gl_{n+1}(K)$, and let us consider 2-reducible, semisimple monoids M with unit group G. Let $\Delta = \{\alpha_1, \ldots, \alpha_n\}$ be the set of simple roots of G, and $\{\omega_1, \ldots, \omega_n\}$ the set of fumdamental, dominant weights. Then it is well known that, for $i = 0, \ldots, n-1$,

$$(n+1)\omega_{i+1} = (n-i)\alpha_1 + 2(n-i)\alpha_2 + \cdots + (i+1)(n-i)\alpha_{i+1} + \cdots + (i+1)\alpha_n.$$

For convenience, we let

$$x_{i+1} = (n+1)\omega_{i+1}.$$

Let M be a 2-reducible, semisimple monoid with unit group G and assume that $I_+ = \Delta\backslash\{\alpha_1\}$, $I_- = \Delta\backslash\{\alpha_{i+1}\}$. The polytope $\mathcal{P}$ of M is the convex hull of the W-orbit of $\{x, y\} \subseteq X(T_0) \otimes \mathbb{Q}^+$. Hence x is a rational multiple of x_1, and y is a rational multiple of x_{i+1}. Without loss of generality, $x = x_1$ and $y = rx_{i+1}$ for some $r > 0$. By the results of Theorem 7.28, we need to calculate

$$x - y = \sum_{i=1}^{n} r_i \alpha_i.$$

But that is elementary, and we obtain

i) $r_j = n - j + 1 - j(r(n-i))$ if $j \leq i$
ii) $r_j = (1-(i+1)r)(n-j+1)$ if $j > i$.

By Corollary 7.27, we must have

i) $n - r(n-i) > 0$, and
ii) $(1-(i+1)r) < 0$.

Hence

$$1/(i+1) \ < \ r \ < \ n/(n-i).$$

For certain special values of r, r_j can be zero. These values are

$$r = (n-j)/(j+1)(n-i).$$

In any case, it is an elementary calculation. We summarize our results as follows.

Theorem 7.29. *Let M be a 2-reducible, semisimple monoid with unit group $Gl_{n+1}(K)$, and assume that $I_+ = \Delta\backslash\{\alpha_1\}$, $I_- = \Delta\backslash\{\alpha_{i+1}\}$. Write $x = x_1$, $y = rx_{i+1}$ as above. Then*

a) $1/(i+1) < r < l/(l-i)$;
b) if $1 \leq j \leq i-1$ *and* $r = (n-j)/(j+1)(n-i)$ *then*
$\Delta_+ = \{\alpha_1, \ldots, \alpha_j\}$
$\Delta_- = \{\alpha_{j+2}, \ldots, \alpha_n\}$;
c) if $0 \leq j \leq i-1$ *and* $(n-j-1)/(j+2)(n-i) < r < (n-j)/(j-i)(n-i)$ *then*
$\Delta_+ = \{\alpha_1, \ldots, \alpha_{j+1}\}$
$\Delta_- = \{\alpha_{j+2}, \ldots, \alpha_n\}$.

It is now possible to calculate Λ and λ in each case using Theorem 7.23. The details are left to the reader.

7.5.5 Monoids with $I_+ = \phi$ and $I_- = \phi$

It is easy to characterize the pairs (I_+, I_-) that can actually occur as $(\lambda_*(e_+), \lambda_*(e_-))$ for some 2-reducible semisimple monoid M with $\Lambda_1 = \{e_+, e_-\}$. Indeed, let $A, B \subseteq \Delta$ be any two proper subsets. Then $(A, B) = (I_+, I_-)$ for some semisimple, 2-reducible monoid M if and only if either

i) $A \neq B$, or else
ii) $A = B$ and $|\Delta \backslash A| \geq 2$.

In particular, $I_+ = I_- = \phi$ is possible; in fact generic. Notice that this is equivalent to $\{x, y\}$ being a subset of $\mathcal{C}^0$, the interior of $\mathcal{C}$.

Theorem 7.30. *The following are equivalent:*

a) there exists a 2-reducible, semisimple monoid M with $I_+ = I_- = \phi$ and $(\Delta_+, \Delta_-) = (U, V)$;
b) $U \neq \phi, V \neq \phi$ and $U \cap V = \phi$.

Proof. Obviously, a) implies b). So assume that $U, V \subseteq \Delta$ satisfy b). Define

$$\delta = \sum_{\alpha \in U} \alpha - \sum_{\beta \in V} \beta.$$

It is then easy to find $x, y \in \mathcal{C}^0$ so that $x - y = \delta$. Then apply Theorem 7.28.

7.5.6 ($\mathcal{J}$,σ)-irreducible Monoids Revisited

In this section we use the results of Theorems 7.23 and 7.28 to study the orbit structure of certain reductive monoids M with involution $\sigma : M \to M$.

Definition 7.31. *Let M be an reductive monoid with zero, and suppose that $\sigma : M \to M$ is a bijective morphism of algebraic monoids. We say that (M, σ) is* ($\mathcal{J}$,σ)-irreducible *if the map induced by σ is transitive on the set of minimal $G \times G$-orbits of $M \backslash \{0\}$.*

($\mathcal{J}$,σ)-irreducible monoids were studied systematically by Z. Li and the other authors of [51, 52, 53]. In all cases, except those that contain D_4 as a component, σ^2 induces the identity morphism on the set of $G \times G$-orbits of M. In such cases, M is a 2-reducible monoid precisely when $M \backslash \{0\}$ has exactly two minimal $G \times G$-orbits and σ exchanges these orbits. In this section, we discuss several examples where M is 2-reducible and semisimple, and σ is actually an automorphism of M of order two. The purpose of Theorems 7.23 and 7.28 is to identify the minimal information (i.e. Δ_+ and Δ_-) needed to get the type map of M.

Example 7.32. Let M be a 2-reducible, semisimple monoid with unit group $Gl_6(K)$. Assume that there is an automorphism $\sigma : M \to M$ such that $\sigma^2 = id$ and $\sigma | Gl_6(K)$ is transpose-inverse.

Let $F = \{\lambda_1, \ldots, \lambda_5\}$ be the set of fundamental dominant weights of $Sl_6(K)$. Then σ induces the following involution σ^* on F:

$$\sigma^*(\lambda_i) = \lambda_{6-i}.$$

From Table 2 on page 295 of [69] we obtain

$$\lambda_1 - \lambda_5 = \frac{1}{6}(4\alpha_1 + 2\alpha_2 - 2\alpha_4 - 4\alpha_5),$$

and

$$\lambda_2 - \lambda_4 = \frac{1}{6}(2\alpha_1 + 4\alpha_2 - 4\alpha_4 - 2\alpha_5).$$

Now any 2-reducible, semisimple monoid M has a representaion $\rho : M \to End(V \oplus W)$, as in Proposition 7.19. If V is the irreducible M-module with highest weight $\lambda \in X(T)_+$, then W is the irreducible M-module with highest weight $\sigma^*(\lambda) \neq \lambda$. Write

$$\lambda = a_1\lambda_1 + a_2\lambda_2 + a_3\lambda_3 + a_4\lambda_4 + a_5\lambda_5$$

where $a_i \geq 0$, and either $a_1 \neq a_5$ or else $a_2 \neq a_4$ (so that $\sigma^*(\lambda) \neq \lambda$). In any case,

$$\begin{aligned}\lambda - \sigma^*(\lambda) = &\frac{1}{6}([4(a_1 - a_5) + 2(a_2 - a_4)]\alpha_1 + [2(a_1 - a_5) + 4(a_2 - a_4)]\alpha_2) \\ &- \frac{1}{6}([2(a_1 - a_5) + 4(a_2 - a_4)]\alpha_4 + [4(a_1 - a_5) + 2(a_2 - a_4)]\alpha_5).\end{aligned}$$

Now

$$\begin{aligned}I_+ &= \{\alpha_i \mid a_i \neq 0\} \\ I_- &= \{\alpha_i \mid a_{6-i} \neq 0\}.\end{aligned}$$

Notice that in all cases $\Delta_- = \{\alpha_{6-i} \mid \alpha_i \in \Delta_+\}$, while $\alpha_3 \notin \Delta_+ \sqcup \Delta_-$. So it suffices to calculate the possibilities for Δ_+ in terms of λ.

1. $\Delta_+ = \{\alpha_1, \alpha_2\}$ if $2(a_1 - a_5) + (a_2 - a_4) > 0$ and $(a_1 - a_5) + 2(a_2 - a_4) > 0$.
2. $\Delta_+ = \{\alpha_1, \alpha_4\}$ if $2(a_1 - a_5) + (a_2 - a_4) > 0$ and $(a_1 - a_5) + 2(a_2 - a_4) < 0$.
3. $\Delta_+ = \{\alpha_1\}$ if $2(a_1 - a_5) + (a_2 - a_4) > 0$ and $(a_1 - a_5) + 2(a_2 - a_4) = 0$.
4. $\Delta_+ = \{\alpha_2\}$ if $2(a_1 - a_5) + (a_2 - a_4) = 0$ and $(a_1 - a_5) + 2(a_2 - a_4) > 0$.

All other feasible data are obtained by reversing the rôles of λ and $\sigma^*(\lambda)$. But we obtain no new monoids. The potential cases with $\Delta_+ = \{\alpha_1, \alpha_5\}$ or $\{\alpha_2, \alpha_4\}$ are not possible. Also, any situation where $|\Delta_+| \geq 3$ is not possible.

We see from Theorems 7.23 and 7.28 that the type map of M is now determined in each case.

Example 7.33. Let M be a 2-reducible, semisimple monoid with unit group $K^*SO_{2n}(K) \subseteq Gl_{2n}(K)$. Assume that there is an automorphism $\sigma : M \to M$ such that $\sigma^2 = id$ and $\sigma|SO_{2n}(K)$ is transpose-inverse.

Let $F = \{\lambda_1, \ldots, \lambda_{n-2}, \lambda_{n-1}, \lambda_n\}$ be the set of fundamental dominant weights of $SO_{2n}(K)$. Then σ induces the following involution σ^* on F:

$$\sigma^*(\lambda_i) = \lambda_i \text{ if } i \leq n-2,$$

and

$$\sigma^*(\lambda_{n-1}) = \lambda_n, \ \sigma^*(\lambda_n) = \lambda_{n-1}.$$

From Table 2 on page 296 of [69] we obtain

$$\lambda_m - \lambda_{n-1} = \frac{1}{2}(-\alpha_{n-1} + \alpha_n).$$

As in the previous example, any 2-reducible, semisimple monoid M has a representaion $\rho : M \to End(V \oplus W)$, according to Proposition 7.19. If V is the irreducible M-module with highest weight $\lambda \in X(\overline{T})_+$, then W is the irreducible M-module with highest weight $\sigma^*(\lambda) \neq \lambda$. Write

$$\lambda = a_1\lambda_1 + a_2\lambda_2 + \cdots + a_{n-2}\lambda_{n-2} + a_{n-1}\lambda_{n-1} + a_n\lambda_n$$

where $a_n \neq a_{n-1}$, (so that $\sigma^*(\lambda) \neq \lambda$). Then

$$\lambda - \sigma^*(\lambda) = \frac{a_n - a_{n-1}}{2}(-\alpha_{n-1} + \alpha_n).$$

Now

$$\begin{aligned} I_+ &= \{\alpha_i \mid a_i \neq 0\} \\ I_- &= \{\alpha_i \mid \overline{a_i} \neq 0\}. \end{aligned}$$

where $\overline{a_n} = a_{n-1}$, $\overline{a_{n-1}} = a_n$, and $a_i = a_i$ if $i < n-1$. Notice again that, in all cases, $\Delta_- = \{\overline{\alpha} \mid \alpha \in \Delta_+\}$, and so we only need to consider the possibilities for Δ_+ in terms of λ. There are just two cases:

1. $\Delta_+ = \{\alpha_{n-1}\}$ if $a_{n-1} \geq a_n$;
2. $\Delta_+ = \{\alpha_n\}$ if $a_n \geq a_{n-1}$.

Again we see from Theorems 7.23 and 7.28 how the type map of M is completely determined in each case. The details are left to the reader. Notice that the two cases yield the same monoid M, since σ^* exchanges α_{n-1} and α_n.

7.6 Type Maps in General

It is certain that the combinatorial classification of type maps of all semisimple monoids M is a "dead end" problem. Indeed, it appears to include the combinatorial classification of all rational polytopes as a proper subproblem. But there are still some interesting questions here. It is clear that the type map is the combinatorial glue that makes the monoid structure possible. But it may also be (as it is for the case of two 0-minimal $\mathcal{J}$-classes in Theorem 7.28) an important combinatorial manifestation of the classification data of reductive monoids.

In this section, we speculate on the likelihood that the set of isomorphism classes of reductive monoids may have the structure of a union of rational polyhedral cones, similar to the data one obtains from a non affine torus embedding. Each face appears to represent the set of isomorphism classes of monoids with the same (fixed) type map. The order relation between these faces should represent a particular combinatorial degeneracy of that type map. This speculation leads us to a number of interesting results about the geometric underpinnings of type maps.

Let G be a semisimple algebraic group with maximal torus T. Let $X(T)$ be the set of characters of T and let $\Delta \subseteq X(T)$ be the set of simple roots. As usual, let

$$\mathcal{C} = \{x \in X(T) \otimes \mathbb{Q} \mid <\alpha, x> \geq 0 \text{ for all } \alpha \in \Delta\}$$

be the Weyl chamber of $E = X(T) \otimes \mathbb{Q}$ associated with Δ.

Definition 7.34. *If $x_1, \ldots, x_n \in \mathcal{C}$, we say that $\{x_1, \ldots, x_n\}$ is* stable *if, for each $i \neq j$,*

$$x_i - x_j = \sum_{\alpha \in \Delta} r_\alpha \alpha$$

has the property that $r_\alpha < 0$ for some $\alpha \in \Delta$, and $r_\beta > 0$ for some $\beta \in \Delta$.

Conjecture 7.35. The following are equivalent for $\{x_1, \ldots, x_n\} \subseteq \mathcal{C}$:

a) $\{x_1, \ldots, x_n\}$ is stable,
b) Each x_i is an extreme point of the convex hull of $\{w(x_i) \mid w \in W,\ i = 1, \ldots, n\}$.

Question 7.36. Write $\Lambda_1 = \{x_1, \ldots, x_n\} \subseteq \mathcal{C}$. Define, for $\alpha \in \Delta$,

$$r_\alpha : \Lambda_1 \times \Lambda_1 \to \mathbb{Q}$$

by the rule

$$x - y = \sum_{\alpha \in \Delta} r_\alpha(x, y)\alpha.$$

Then

a) $r_\alpha(x,y) + r_\alpha(y,z) = r_\alpha(x,z)$;
b) $r_\alpha(x,y) = -r_\alpha(y,x)$;
c) if $x \neq y$ then there exists $\alpha, \beta \in \Delta$ such that $r_\alpha(x,y) > 0$ and $r_\beta(x,y) < 0$.

Does any collection $\{r_\alpha : \Lambda_1 \times \Lambda_1 \to \mathbb{Q}\}$ satisfyting a), b) and c) come from a subset $\Lambda_1 \subseteq \mathcal{C}$? If not, is there an interpretation?

Given $\Lambda_1, \Lambda_1' \subseteq \mathcal{C}$ as in Definition 7.34 we say that Λ_1 and Λ_1' have the same **shape** if there is a bijection $\rho : \Lambda_1 \to \Lambda_1'$ such that

a) $W_x = W_{\rho(x)}$ for each $x \in \Lambda_1$
b) $r_\alpha(x,y) = 0$ if and only if $r_\alpha(\rho(x), \rho(y)) = 0$
c) $r_\alpha(x,y) > 0$ if and only if $r_\alpha(\rho(x), \rho(y)) > 0$.

We do not claim here that, if Λ_1 and Λ_1' have the same shape, then they come from monoids with the same type map. This does not seem to be true, although we do not yet have any revealing examples.

Proposition 7.37. *Assuming the above conjecture is true, the bijection $\rho : \Lambda_1 \to \Lambda_1'$ is unique if it exists.*

Proof. Suppose that there there are two, say $\rho : \Lambda_1 \to \Lambda_1'$ and $\sigma : \Lambda_1 \to \Lambda_1'$. Then let $\psi = \rho^{-1} \circ \sigma : \Lambda_1 \to \Lambda_1$. Notice that ψ satisfies a), b) and c) above. By Conjecture 7.35, there exists $\alpha \in \Delta$ such that

$$r_\alpha(x, \psi(x)) > 0.$$

Thus,

$$r_\alpha(\psi(x), \psi^2(x)) > 0,$$

$$\vdots$$

$$r_\alpha(\psi^{n-1}(x), \psi^n(x)).$$

Also, where we assume that $\psi^n(x) = x$. In any case,

$$\sum_{i=1}^{n} r_\alpha(\psi^{i-1}(x), \psi^i(x)) > 0.$$

However,

$$\begin{aligned}
0 &= (x - \psi(x)) + (\psi(x) - \psi^2(x)) + \cdots + (\psi^{n-1}(x) - \psi^n(x)) \\
&= \sum_{\alpha \in \Delta} r_\alpha(x, \psi(x))\alpha + \cdots + \sum_{\alpha \in \Delta} r_\alpha(\psi^{n-1}(x), \psi^n(x))\alpha \\
&= \sum_{\alpha \in \Delta} (\sum_{i=1}^{n} r_\alpha(\psi^{i-1}(x), \psi^i(x)))\alpha.
\end{aligned}$$

Thus, $\sum_{i=1}^{n} r_\alpha(\psi^{i-1}(x), \psi^i(x)) = 0$, since $\Delta \subseteq E$ is a $\mathbb{Q}$-basis. This contradiction finishes the proof.

We conclude that, if Λ_1 and Λ_1' have the same shape, then we can add them as follows.
Let $\rho : \Lambda_1 \to \Lambda_1'$ be the unique bijection that evidences Λ_1 and Λ_1' of the same shape. Then define the **sum** of Λ_1 and Λ_1' as

$$\Lambda_1'' = \{x + \rho(x) \mid x \in \Lambda_1\}.$$

Proposition 7.38. *Λ_1'' has the same shape as Λ_1.*

Proof. Define $\psi : \Lambda_1 \to \Lambda_1''$ by $\psi(x) = x + \rho(x)$. Now for $x \in \Lambda_1$,

$$x, \rho(x) \in (E^{W_x} \cap \mathcal{C})^0,$$

which is closed under addition. Hence $x + \rho(x) \in (E^{W_x} \cap \mathcal{C})^0$ as well, and thus $W_x = W_{x+\rho(x)}$, since

$$(E^{W_x} \cap \mathcal{C})^0 = \{y \in \mathcal{C} \mid W_y = W_x\}.$$

To finish the proof, notice that

$$x + \rho(x) - (y + \rho(y)) = \sum_{\alpha \in \Delta} (r_\alpha(x, y) + r_\alpha(\rho(x), \rho(y))\alpha.$$

Hence $r_\alpha(x, y) = 0$ implies that $r_\alpha(\rho(x), \rho(y)) = 0$, which implies that $r_\alpha(x + \rho(x), y + \rho(y)) = 0$. Similarly for $>$ and $<$. Thus Λ_1 and Λ_1'' have the same shape.

One can pose the dual problem using Λ^1 and λ_*. One can determine the type map in terms of colors (λ_*) and divisors (λ^1) (see § 5.3.3). One might then be able to define the addition of polytopes and cross section lattices (in that setup) in terms of the associated valuations comming from Λ^1.

7.7 Exercises

7.7.1 The Cross Section Lattice

1. Let M be reductive, and let $\Lambda \subseteq E(\overline{T})$ be a cross section lattice. Prove that the number of maximal chains in $E(\overline{T})$ is equal to the number of maximal chains of Λ times the order of the Weyl group.
2. One defines a reductive monoid M, with zero, to be $\mathfrak{J}_i$*-irreducible* if $|\Lambda_j| = 1$ for all $j \leq i$. Prove that
   i) M is $\mathfrak{J}_2$-irreducible if and only if $J_0 = S\backslash\{s\}$ for some $s \in S$
   ii) M is $\mathfrak{J}_3$-irreducible if and only if $J_0 = S\backslash\{s\}$ where s corresponds to an end node on the Dynkin diagram of G.
3. One can use Theorem 7.12 also to characterize other classes of $\mathfrak{J}$-irreducible monoids.

i) M is $\mathfrak{J}$-simple if and only if S is connected and M is either $\mathfrak{J}_2$-irreducible or $S\backslash J_0 = \{s, t\}$ where $st \neq ts$. Here, we say a $\mathfrak{J}$-irreducible monoid is *$\mathfrak{J}$-simple* if $\lambda^*(e)$ is a connected subset of the Dynkin diagram for each $e \in \Lambda$.

ii) $\Lambda(M)$ is a distributive lattice if and only if $S\backslash J_0$ is connected.

4. Let M be reductive, and let Λ be a cross section lattice of M. Prove that there is a one-to-one correspondence between the set of two-sided ideals of M and the set of poset ideals of Λ.

7.7.2 Idempotents

1. Let $\psi : M \to N$ be a finite dominant morphism of irreducible algebraic monoids.
   a) Prove that $\mathcal{U}(\psi) : \mathcal{U}(M) \to \mathcal{U}(N)$ is bijective.
   b) Prove that $E(\psi) : E(M) \to E(N)$ is bijective.
2. Let M be irreducible with unit group G and maximal torus T. Let B be a Borel subgroup containing T. Let α be a positive root, and consider $U_\alpha \subseteq B, e \in E(\overline{T})$.
   a) Prove the following are equivalent:
      i) $eU_\alpha = U_\alpha e$,
      ii) $s_\alpha e = es_\alpha$.
   b) Prove the following are equivalent:
      i) $eU_\alpha = U_\alpha e \neq \{e\}$,
      ii) $s_\alpha e = es_\alpha \neq e$.
3. Let M be reductive with $e \leq f \leq g$. Assume that $e, f, g \in E(\overline{T})$. As usual, let $S = \{s_\alpha \in W \mid \alpha \in \Delta\}$, and identify S with the set of nodes on the Dynkin diagram. Prove that each connected component of $\lambda^*(f)$ is contained in either $\lambda^*(e)$ or $\lambda^*(g)$.

8

The Analogue of the Bruhat Decomposition

Recall from § 2.2.3 that any reductive group G has a Bruhat decomposition:

$$G = \bigsqcup_{w \in W} BwB$$

where $W = N_G(T)/T$. In this chapter we extend this result to reductive monoids. Instead of W we use

$$R = \overline{N_G(T)}/T$$

where $\overline{N_G(T)} \subseteq M$ is the Zariski closure of $N_G(T)$ in M. Since $xT = Tx$ for each $x \in \overline{N_G(T)}$, R is a monoid, but much more is true. It turns out that the following are true (and will be explained in this chapter).

a) R is a finite inverse semigroup with unit group W.
b) $M = \bigsqcup_{x \in R} BxB$, a disjoint union.
c) $sBx \subseteq BxB \sqcup BsxB$ if s is a simple involution and $x \in R$ (Tits' axiom).
d) There is a canonical length function $\ell : R \to \mathbb{N}$.
e) If we define $x \leq y$ to mean $BxB \subseteq \overline{ByB}$, we can determine $(R, \leq)$ in terms of $(W, \leq)$ and the cross section lattice.
f) If $x \leq y$ and $\ell(x) = \ell(y) - 1$, then either $x \in R^+yR^+$ or else y is obtained from x by an elementary "interchange" exactly as in the case of a Coxeter group (Pennell's Theorem).
vii) There is a combinatorial description of $(R, \leq)$ in the case of $M = M_n(K)$.

8.1 The Renner Monoid R

A monoid S is called **inverse** if for each $x \in S$ there exists a unique $x^* \in S$ such that

$$x^*xx^* = x^*,$$
$$xx^*x = x,$$

and

$$(x^*)^* = x\ .$$

The standard example here is

$$R_n = \left\{ A \in M_n(K) \,\middle|\, \begin{array}{l} a_{ij} = 0 \text{ or } 1 \\ \Sigma_i a_{ij} \le 1 \text{ for all } j \\ \Sigma_j a_{ij} \le 1 \text{ for all } i \end{array} \right\}$$

where $A^* = A^t$. R_n is isomorphic to the semigroup of partial one-to-one functions on a set of n elements.

Proposition 8.1. *Let $R = \overline{N_G(T)}/T$. Then R is a finite inverse monoid with unit group $W = N_G(T)/T$ and idempotent set $E(R) = E(\overline{T})$.*

Proof. Note that $\overline{N_G(T)} = N_G(T)\overline{T}$ since the latter is closed in $\overline{N_G(T)}$. Thus $R = WE$ where $E \subseteq R$ is the image of $E(\overline{T})$ in R.

Now assume that $xTxT = xT$. Then $xT = x^2T$ so that $x^2 = tx$ for some $t \in T$. Thus, $x^n = t^{n-1}x$ for all $n > 0$. But $x^m \in \overline{T}$ for $m = |W|$. Hence $x = t^{1-m}x^m \in \overline{T}$ and so $t^{-1}x \in \overline{T}$, while $\pi(t^{-1}x) = \pi(x)$ where $\pi : \overline{N_G(T)} \longrightarrow R$ is the quotient map. But $(t^{-1}x)^2 = t^{-2}x^2 = t^{-2}tx = t^{-1}x$. Hence $E(R) = E(\overline{T})$.

Now $W \subseteq R^*$, the unit group of R. If $x \in R$ then $x = we$ for some $w \in W$ and $e \in E(\overline{T})$. If $e \neq 1$ then $E(\overline{Z}) \longrightarrow E(\overline{Z})$, $f \longmapsto ef$ is not 1-1. Hence x cannot be a unit.

But now R is a regular monoid with commutative idempotent set. Thus R is an inverse semigroup by Theorem 1.17 of [15].

Any reductive group G has an involution $\tau : G \to G$ such that

$$\tau^2 = id,$$
$$\tau(xy) = \tau(y)\tau(x) \text{ for all } x, y \in G,$$

and

$$\tau(x) = x \text{ for all } x \in \overline{T}.$$

Using Theorem 5.2, we can extend τ to an involution $\tau : M \to M$ for any normal, reductive monoid M with unit group G. It then follows easily that τ induces a map

$$\overline{\tau} : R \to R\ .$$

One checks that $\overline{\tau}(x) = x^*$ for all $x \in R$, since $\overline{\tau}(e) = e$ for all $e \in E(R)$, while $\overline{\tau}(w) = w^{-1}$ for all $w \in W$.

8.2 The Analogue of the Tits System

In this section we establish the fundamental results about the $B \times B$ orbits of M. We start with a special case, and then use it to build the general case.

Proposition 8.2. *Let M be a reductive monoid with unit group $G = S\ell_2 \times K^*$, $G\ell_2$ or $PG\ell_2 \times K^*$, and zero element $0 \in M$.*

a) $R = \{1, s, e_1, e_2, n_1, n_2, 0\}$.

b) $M = \bigsqcup_{r \in R} BrB$.

c) In the case $M = M_2(K)$,

$$R = R_2 = \left\{ \begin{pmatrix} 1 & 0 \\ 0 & 1 \end{pmatrix}, \begin{pmatrix} 0 & 1 \\ 1 & 0 \end{pmatrix}, \begin{pmatrix} 1 & 0 \\ 0 & 0 \end{pmatrix}, \begin{pmatrix} 0 & 0 \\ 0 & 1 \end{pmatrix}, \begin{pmatrix} 0 & 1 \\ 0 & 0 \end{pmatrix}, \begin{pmatrix} 0 & 0 \\ 1 & 0 \end{pmatrix}, \begin{pmatrix} 0 & 0 \\ 0 & 0 \end{pmatrix} \right\} .$$

d) In all cases, $R \cong R_2$ as semigroups, with the orders given.

Proof. The proof of c) is a straightforward calculation. Also b) holds for $M_2(K)$ by a simple calculation. So let M be as assumed. By Section 6 of [103], there exists a diagram

$$\begin{array}{ccc} M' & \xrightarrow{\alpha} & M_2(K) \\ {\scriptstyle\beta}\big\downarrow & & \\ M & & \end{array}$$

where α and β are finite and dominant morphisms of algebraic monoids. It is then easily checked that α and β both induce isomorphisms

$$\begin{array}{ccc} R' & \underset{\cong}{\xrightarrow{\overline{\alpha}}} & R \\ {\scriptstyle\overline{\beta}}\big\downarrow{\scriptstyle\cong} & & \\ R_2 & & \end{array}$$

This concludes the proof.

Proposition 8.3. *Suppose that M is reductive with unit group $G \cong Sl_2 \times K^*$, $Gl_2(K)$ or $PGl_2 \times K^*$, but M does not have a zero. Assume also that G is not equal to M. Then $R = \{s, 1, e, x = se = es\}$. R is isomorphic (with the ordering given) to*

$$\left\{ \left(\begin{pmatrix} 0 & i \\ -i & 0 \end{pmatrix}, 1 \right), \left(\begin{pmatrix} 1 & 0 \\ 0 & 1 \end{pmatrix}, 1 \right), \left(\begin{pmatrix} 0 & i \\ -i & 0 \end{pmatrix}, 0 \right), \left(\begin{pmatrix} 1 & 0 \\ 0 & 1 \end{pmatrix}, 0 \right) \right\} .$$

Proof. There exists a finite dominant morphism $\pi : Sl_2(K) \times K \to M$. Then the result follows for M as in Proposition 8.2.

Proposition 8.4. *Let M be any reductive monoid such that $G = G(M)$ has semisimple rank one. Let T, R and B have the usual meanings, so that $W = \{1, s\} \subseteq R$. Then for any $x \in R$*

$$sBx \subseteq BxB \cup BsxB\ .$$

Proof. Now $R = WE = E \cup sE$ by Proposition 8.1. So let $x = se \in R$. Let $G' = (G, G)$ and choose $K^* \subseteq T$ such that $e \in \overline{K^*}$. Then $M' = \overline{K^*G'}$ is of the form discussed in 8.2 or 8.3. But also $x \in R'$. Then the result follows for M' by a simple calculation with $M_2(K)$ and $S\ell_2(K) \times K$. To get the result for M notice that, if $B' \subseteq G'$ is a Borel subgroup, then $B = B'Z(G)^\circ$ is a Borel subgroup of G. Hence

$$\begin{aligned} sBx &= sB'Sx = sB'xs \\ &\subseteq (B'xB' \cup B'sxB')S \\ &= B'xB'S \cup B'sxB'S \\ &= BxB \cup BsxB. \end{aligned}$$

We now explain the general case. So let M be reductive with T, $\overline{T}$, R, B, Φ etc. as usual. For $\alpha \in \Phi$ let $s = s_\alpha \in S$ and let $T^s = \{t \in T \mid st = ts\}$. Let

$$\begin{aligned} T_\alpha &= (T^s)^0, \\ Z_\alpha &= C_G(T_\alpha) \end{aligned}$$

and

$$G_\alpha = (Z_\alpha, Z_\alpha).$$

Then $G_\alpha = S\ell_2(K)$ or $PG\ell_2(K)$.

Lemma 8.5. *Let $\alpha \in \Phi$, $x \in R$. Then*

$$Z_\alpha xB = U_\alpha xB \cup U_\alpha sxB$$

where $s = s_\alpha \in W$.

Proof. One can check that either (i) $U_\alpha xB = xB$, or else (ii) $U_{-\alpha}xB = xB$. See Lemma 5.2 of [105]. If (i) holds, then

$$\begin{aligned} Z_\alpha xB &= T_\alpha G_\alpha xB \\ &= T_\alpha(B_\alpha \cup B_\alpha sB_\alpha)xB, \text{ where } B_\alpha = B \cap G_\alpha, \\ &= TU_\alpha xB \cup TU_\alpha sB_\alpha xB \\ &= U_\alpha xB \cup U_\alpha sTxB, \text{ by (i)}, \\ &= U_\alpha xB \cup U_\alpha sxB, \end{aligned}$$

using standard properties relating R, T and B. If (ii) holds, then we obtain in the same way

$$Z_\alpha xB = U_{-\alpha}xB \cup U_{-\alpha}sxB \ .$$

But $s(Z_\alpha xB) = Z_\alpha xB$ since $s \in Z_\alpha$. Hence

$$\begin{aligned} Z_\alpha xB &= sU_{-\alpha}xB \cup sU_{-\alpha}sxB \\ &= U_\alpha sxB \cup U_\alpha xB. \end{aligned}$$

Theorem 8.6. *Let $x \in R$ and $s \in S$, where $s = s_\alpha$, $\alpha \in \Delta$, the set of simple roots relative to B. Then*

$$sBx \subseteq BxB \cup BsxB \ .$$

Proof. Let $V = B_u \cap sB_us$. Then $VU_\alpha = U_\alpha V = B_u$, and $sV = Vs$. Hence

$$\begin{aligned} sBx \subseteq sBxB &= sVU_\alpha TxB \\ &= sVU_\alpha xB \\ &\subseteq VZ_\alpha xB, \text{ since } sU_\alpha \subseteq Z_\alpha, \\ &= V(U_\alpha xB \cup U_\alpha sxB) \\ &= BxB \cup BsxB \ . \end{aligned}$$

One could deduce also that

$$xBs \subseteq BxB \cup BxsB \ .$$

We could do this directly, or we could use the involution of § 8.1 applied to the result of Theorem 8.6.

Proposition 8.7. *a) Let $e \in E(R)$, $x \in R$ and suppose that $BeB = BxB$. Then $e = x$.*
b) Let $x, y \in R$ and $BxB = ByB$. Then $x = y$.

Proof. We start with a). Write $x = e_1w = we_2$ where $w \in W$ and $e_1, e_2 \in E(R)$. Now $eBxBe = eBeBe$ and so $exe = ue = eu$ for some $u \in B$. Thus, $ee_1 = ee_2 = e$. But also $C_B(e_1)xC_B(e_2) = e_1BxBe_2 = e_1BeBe_2$. Hence $e_1ee_2 = u_1xu_2$ for some $u_i \in C_B(e_i)$. Thus $ee_1 = e_1$ and $ee_2 = e_2$. One concludes that

$$\begin{aligned} x = ew = we&, \text{ for some } w \in W \\ = eu = ue&, \text{ for some } u \in B \ . \end{aligned}$$

But then $x, e \in R(eMe)$ while

$$eBe\ e\ eBe = eBe\ x\ eBe \ .$$

So by induction (on $dimM$), $x = e$ in R since eMe is a reductive monoid with unit group $eC_G(e)$.

For b), recall first that W is generated by $S = \{s \in W \mid s = s_\alpha \text{ for some } \alpha \in \Delta\}$. Define for $x \in R$

$$\ell(x) = min\{\ell \mid x = \rho_1 \cdot \dots \cdot \rho_\ell e, \text{ for some } e \in ER,\ s_i \in S\}\ .$$

This is not the usual length function on R (as in Definition 8.17 below), but it is useful in this situation. Our proof, imitating Theorem 29.2 of [40], proceeds by induction on $m = \ell(x) \leq \ell(y)$; the case $m = 0$ being part (a) above. So assume that $1 \leq \ell(x) \leq \ell(y)$ and $BxB = ByB$. Write $x = sx^*$, $s \in S$ and $\ell(x^*) = \ell(x) - 1$. Then

$$xB = sx^*B \subseteq ByB\ .$$

But then

$$x^*B \subseteq sByB \subseteq ByB \cup BsyB \text{ by } 8.6.$$

Hence, either $x^*B \subseteq ByB$ or else $x^*B \subseteq BsyB$.

The former case is not possible by the induction hypothesis since $\ell(x^*) < \ell(x) \leq \ell(y)$. But in the latter case, we must have $x^* = sy$, again by induction. But then $x = sx^* = s^2y = y$.

Theorem 8.8. $M = \bigsqcup_{x \in R} BxB$ *a disjont union.*

Proof. We know that the union is disjoint from Proposition 8.7. Let $L = \bigsqcup_{x \in R} BxB$. We show that $GLG = L$. Now $G = \bigsqcup_{w \in W} BwB$, and so it suffices to show that $BwBL \subseteq L$, for each $w \in W$. Write $w = vs$ where $s \in S$ and $\ell(w) = \ell(v) + 1$. $\ell(\cdot)$ is the usual length function on W. Now

$$\begin{aligned} BwBL &= BwB(\sqcup BxB) \\ &= BvB(\sqcup sBxB) \\ &\subseteq BvB(\sqcup(BxB \cup BsxB)), \text{ by Theorem 8.6,} \\ &= BvBL \\ &\subseteq L \text{ by induction on length.} \end{aligned}$$

Similarly, $LG = L$ by the comment following Theorem 8.6. But $GE(R)G = M$ and so $L = M$.

8.3 Row Reduced Echelon Form

The most important technique in linear algebra is the Gauss-Jordan algorithm: given any $m \times n$ matrix A we obtain a matrix $GJ(A)$, in row reduced echelon form, by a procedure known as row-reduction. In the case $m = n$, the procedure solves the orbit classification problem for the action

$$G\ell_n(K) \times M_n(K) \longrightarrow M_n(K)\ ,$$

$$(g, A) \longmapsto gA\ .$$

In this section we discuss this problem in general. Let M be reductive with unit group G, and consider the action

$$\begin{aligned} G \times M &\longrightarrow M \ , \\ (g, x) &\longmapsto gx \ . \end{aligned}$$

We want to define a nice subset $X \subseteq M$ such that for all $x \in M$,

$$Gx \cap X \text{ contains exactly one element.}$$

In this section we omit some details of the proofs. Those details can be found in Section 9 of [105].

So let M, G, B, T, S, W, R and Λ be as usual, with $\Lambda = \{e \in E(\overline{T}) \mid Be \subseteq eB\}$. We first solve the problem in R.

Proposition 8.9. *Let $f \in E(\overline{T})$. Choose $w \in W$, of minimal length, such that $wfw^{-1} = e \in \Lambda$. Then $Bwf \subseteq wfB$. Furthermore, $ew = wf \in Wf$ is the unique element of Wf with this property.*

Proof. By Lemma 9.2 of [105], $Bew \subseteq ewB$ iff $U_\alpha \subseteq wBw^{-1}$ whenever $\alpha \in \Delta$ and $U_\alpha e = eU_\alpha \neq \{e\}$. Suppose this criterion does not hold: say $U_\alpha \nsubseteq wBw^{-1}$ with $U_\alpha e = eU_\alpha \neq \{e\}$. Then $\ell(s_\alpha w) < \ell(w)$ and yet $(s_\alpha w)^{-1} e (s_\alpha w) = w^{-1}ew = f$: a contradiction.

Now suppose that vf, $wf \in Wf$, and both satisfy the property. Then $vfv^{-1} = wfw^{-1} = e \in \Lambda$. Hence $vw^{-1}e = evw^{-1}$. One then checks as in Theorem 9.6 of [105] that $vw^{-1}e = e$.

Definition 8.10. *Let $\mathfrak{GJ} = \{x \in R \mid Bx \subseteq xB\}$ be the set of* Gauss-Jordan elements *of R.*

By Proposition 8.9,

a) $W \cdot \mathfrak{GJ} = R$ and
b) For each $x \in R$, $|Wx \cap \mathfrak{GJ}| = 1$.

It follows easily, also from Proposition 8.9 that, for any $x \in M$, $Gx \cap rB \neq \phi$ for some unique $r \in \mathfrak{GJ}$. This is roughly equivalent to saying that x can be put into row echelon form.

Example 8.11. Let $M = M_n(K)$ with T diagonal and B upper triangular. Then R is identified with the set of $0-1$ matrices $x = (x_{ij})$ such that $\sum_i x_{ij} \leq 1$ for all j and $\sum_j x_{ij} \leq 1$ for all i. For $x \in R$ we obtain that $x \in \mathfrak{GJ}$ iff x is in row-reduced echelon form. For example,

$$x = \begin{pmatrix} 1&0&0&0 \\ 0&0&1&0 \\ 0&0&0&1 \\ 0&0&0&0 \end{pmatrix} \in \mathfrak{GJ} \ .$$

We can identify $\mathcal{GJ}$ with $E(\overline{T})$ using Proposition 8.9.

Given $f \in E(\overline{T})$ there is a unique $r \in Wf \cap \mathcal{GJ}$. In the above example, x corresponds to $\begin{pmatrix} 1&0&0&0 \\ 0&0&0&0 \\ 0&0&1&0 \\ 0&0&0&1 \end{pmatrix} \in E(D_4)$.

We return to our task.

Definition 8.12. *Let M be reductive with B, T, R as usual. An element $x \in M$ is in* reduced form *if*

(a) $x \in rB$ for some $r \in B$;
(b) $xr^* \in \Lambda$ where $r^* \in R$ is the inverse of r.

Said differently, if $r = wf \in \mathcal{GJ}$ and $x = wfu$ with $u \in B_u$, then $fuf = f$.

In the classical situation, (a) says that x is in row echelon form and (b) says (roughly) that it is reduced.

Theorem 8.13. *Let $x \in M$. Then $Gx \cap rB \neq \phi$ for some unique $r \in \mathcal{GJ}$. Furthermore, there is a unique T-orbit in $Gx \cap rB$ consisting of elements in reduced form.*

Proof. (Sketch) The first part has already been proved. So assume that $x \in rB$. Then we can write

$$x = wfy, \text{ where } r = wf, \text{ and } y \in U.$$

But we can write

$$U = U_1 U_2$$

where

$$U_1 = \prod_{\alpha \in A} U_\alpha, \ A = \{\alpha \in \Phi^+ \mid U_\alpha e = eU_\alpha \neq e\}$$

$$U_2 = \prod_{\alpha \in B} U_\alpha, \ B = \{\alpha \in \Phi^+ \mid eU_\alpha e = e\}.$$

Hence

$$\begin{aligned} x &= wfy \\ &= wfuv, \text{ where } u \in U_1, \ v \in U_2, \\ &= u^w fv \,. \end{aligned}$$

One checks that $u^w \in U$ and also that $wfv = (u^w)^{-1}x$ is reduced.

To show uniqueness of wfv up to T-orbit, assume that $x_1, x_2 \in rB$ are both in reduced form with $gx_1 = x_2$. One then shows that $gx_1 = bx_1 = x_2$ for some $b = tu \in B = TU$. It then follows that $ux_1 = x_1$. Hence $x_2 = tx_1$. See Lemma 9.9 of [105].

Corollary 8.14. *Let $x \in M$. Then*

a) there is a unique $e \in E(R)$ such that $GxB = GeB$;
b) $|Gx \cap E(eB)| = 1$.

Proof. Let $z \in Gx \cap rB$ be in reduced form, as guaranteed by Theorem 8.13. We can write $z = wfb$ where $r = wf$ and $b \in B$. One then checks that $\{fb\} = Gx \cap E(eB)$. See Lemma 4.1 of [112].

8.4 The Length Function on R

To define the length function on R, we first need to identify the elements of length zero.

Now $R = \bigsqcup_{e \in \Lambda} WeW$. It turns out that each WeW has a unique minimal element.

Proposition 8.15. *There is a unique element $\nu \in WeW$ such that $B\nu = \nu B$.*

Proof. Let $w \in W$ be the longest element and let $f = wew^{-1}$. By Proposition 8.9, there is a unique $\nu \in Wf$ such that $B\nu \subseteq \nu B$. One checks that $\nu B \subseteq B\nu$ also. See Proposition 1.2 of [112].

Example 8.16. Let $M = M_n(K)$. If $e = \begin{pmatrix} 1 & & & & \\ & \ddots & & & \\ & & 1 & & \\ & & & 0 & \\ & & & & \ddots & \\ & & & & & 0 \end{pmatrix}$ has rank $= i$, then

$\nu = \begin{pmatrix} 0 & 1 & 0 \\ & \ddots & \ddots & 1 \\ 0 & & \ddots & 0 \end{pmatrix}$, also of rank $= i$.

Definition 8.17. *Define*

$$\ell : R \longrightarrow \mathbb{N}$$

$$\text{by} \qquad \ell(r) = \dim(BrB) - \dim(B\nu B)$$

$$\text{where} \qquad \nu \in WrW \text{ is such that } B\nu = \nu B\,.$$

Theorem 8.18. *Let $s \in S \subseteq W$ and $r \in R$. Then*

$$BsBrB \equiv \begin{cases} BrB & \text{if } \ell(sr) = \ell(r) \\ BsrB & \text{if } \ell(sr) = \ell(r) + 1 \\ BsrB \cup BrB & \text{if } \ell(sr) = \ell(r) - 1. \end{cases}$$

Proof. By Theorem 8.6 there are the three possibilities. If $BsBrB = BrB$ then, by Proposition 8.7, $sr = r$. Thus $\ell(sr) = \ell(r)$.

Suppose $BsBrB = BsrB$. So $sBrB \subseteq BsrB$ and thus $\ell(sr) \geq \ell(r)$. But $BsBrB = U_\alpha sBrB$ for some unique $\alpha \in \Delta$. Thus $\ell(sr) \leq \ell(r) + 1$. If $\ell(sr) = \ell(r)$ then we obtain $sBrB = BsrB$, and so $sr = r$. If $BsBrB = BsrB$ and $sr \neq r$ then we obtain $\ell(sr) = \ell(r) + 1$.

Suppose that $BsB\nu B = BrB \cup BsrB$ and $sr \neq r$. Then $BsBrB = (BsB \cup B)rB = P_\alpha rB$. One then gets that

$$BrB = Bs(B \backslash V)rB$$

where $V = R_u(P_\alpha)$. Hence $V \subseteq B$ is the unique closed, normal unipotent subgroup of B such that $sV = Vs$ and $VU_\alpha = U_\alpha V = B$. Hence $BsrB \subseteq \overline{BrB}$ and thus $\ell(sr) < \ell(r)$. One then gets $\ell(sr) = \ell(r) - 1$. See Theorem 1.4 [112] for more details.

Example 8.19. Let $M = M_n(K)$, where K is a finite field. Then $M = \bigsqcup_{x \in R_n} BxB$ where R_n is the standard example of § 8.1. In [127], Solomon proves 8.15 and 8.18 for this M. He also defines a length function that agrees with the one in Definition 8.17. He then defines the analogue of Iwahori's Hecke algebra as follows:

$$H(M, B) = \bigoplus_{x \in R_n} \mathbb{Z} \cdot T_x$$

with multiplication defined by

$$T_s T_x = \begin{cases} qT_x \ , & \ell(sx) = \ell(x) \\ T_{sx} \ , & \ell(sx) = \ell(x) + 1 \\ qT_{sx} + (q-1)T_x \ , & \ell(sx) = \ell(x) - 1 \end{cases}$$

$$T_x T_s = \begin{cases} qT_x \ , & \ell(xs) = \ell(x) \\ T_{xs} \ , & \ell(xs) = \ell(x) + 1 \\ qT_{xs} + (q-1)T_x \ , & \ell(xs) = \ell(x) - 1 \end{cases}$$

$$T_\nu T_x = q^{\ell(x) - \ell(\nu x)} T_{\nu x}$$

$$T_x T_\nu = q^{\ell(x) - \ell(x\nu)} T_{x\nu}$$

where

$$\nu = \begin{pmatrix} 0 & 1 & & 0 \\ & \ddots & \ddots & \\ 0 & & \ddots & 1 \\ & & & 0 \end{pmatrix} .$$

He then proves in Theorem 4.12 of [127] that $H(M,B)$ is a ring, with this definition. The delicate part here is obtaining integral structure constants. One suspects that his approach works in general. Putcha approached this problem systematically in [87].

In [129], Solomon has revisited the representation theory of R_n bringing out a rich blend of combinatorics and algebra.

8.5 Order-Preserving Elements of *R*

In this section we assess the relationship of elements of R to B, and obtain the subset of order preserving elements. The standard example here is $M = M_n(K)$, where R is identified with the set of 01 matrices with at most one nonzero entry in each row or column. An element $r \in R$ is order preserving if the matrix obtained from r by deleting all the zero rows and all the zero columns is an identity matrix. The general definition is as follows.

Definition 8.20. *Let $\mathcal{O} \subseteq R$ be the subset of elements $r \in R$ with the property*

$$rBr^* \subset Brr^*$$

where $r^* \in R$ is the unique element satisfying $rr^*r = r$ and $r^*rr^* = r^*$. Any element with this property is called order preserving.

Lemma 8.21. *a) If $r, s \in O$ then $rs \in \mathcal{O}$.*
b) $W \cap \mathcal{O} = \{1\}$.
c) $E(R) \subseteq \mathcal{O}$.
d) If $r \in \mathcal{O}$ then $rBr^* \subseteq rr^*Brr^*$.

Proof. a) Suppose that $rBr^* \subseteq Brr^*$ and $sBs^* \subseteq Bss^*$. Consider $ss^*r^* \in R$. Write $ss^* = e$ and $r^* = f\sigma^{-1}$. Then $ss^*r^* = ef\sigma^{-1} = fe\sigma^{-1} = f\sigma^{-1}\sigma e\sigma^{-1} = r^*(ss^*)^\sigma$. Now compute:

$$\begin{aligned} rsB(rs)^* &= rsBs^*r^* \subseteq rBss^*r^* \\ &= rBr^*(ss^*)^\sigma \subseteq Brr^*(ss^*)^\sigma \\ &= Brss^*r^* = B(rs)(rs)^*. \end{aligned}$$

b) Suppose that $r \in W \cap O$. Then $r^* = r^{-1}$, and $rBr^* \subseteq Brr^*$ implies $rBr^{-1} = B$.
c) If $e \in E(R)$ then, by Theorems 6.16 and 6.30 of [82], $eBe^* = eBe = eC_B(e)e \subseteq Be = Bee^*$.
d) Write $r = e\sigma$ so that $r^* = \sigma^{-1}e$ and $rr^* = e$. Then $rBr^* \subseteq eM \cap Be$. But if $be \in eM \cap Be$ then $be = ex$, and so $ebe = ex = be$. Hence $be \in eBe$. We conclude finally that $rBr^* \subseteq eM \cap Be \subseteq eBe = rr^*Brr^*$.

Corollary 8.22. *The following are equivalent.*

a) $rBr^* \subseteq Brr^*$.
b) $rBr^* \subseteq rr^*B$.
c) $rBr^* \subseteq rr^*Brr^*$.

Proof. $rr^*Brr^* \subseteq Brr^* \cap rr^*B$.

Recall now the involution $\tau : M \to M$. It has the following properties:

$$\tau|T = id,$$
$$\tau(B) = B^-,$$
$$\tau^2 = id,$$
$$\tau(xy) = \tau(y)\tau(x),$$

and

$$\tau(r) = r^*, \text{ for all } r \in R.$$

Recall also $w \in W$, the *longest* element. It satisfies $wBw^{-1} = B^-$.

Proposition 8.23. *Let* $\mathcal{O}^- = \{r \in R | rB^-r^* \subseteq B^-rr^*\}$ *and let* $\mathcal{O}^w = \{wrw^{-1} \in R | r \in \mathcal{O}\}$. *Then* $\mathcal{O} = \mathcal{O}^- = \mathcal{O}^w$.

Proof. Let $r \in R$. Then $r \in \mathcal{O}$ iff $rBr^* \subseteq Brr^*$ iff $\tau(rBr^*) \subset \tau(Brr^*)$ iff $rB^-r^* \subseteq rr^*B^-$ iff $rB^-r^* \subseteq B^-rr^*$ (by Corollary 8.22) iff $r \in \mathcal{O}^-$. Hence $\mathcal{O} = \mathcal{O}^-$.

Again if $r \in R$, $r \in \mathcal{O}^w$ iff $wrwB(wrw)^* \subseteq Bwrw(wrw)^*$ iff $wrB^-r^*w \subseteq Bwrr^*w$ iff $rB^-r^* \subseteq B^-rr^*$ iff $r \in O^-$. Hence $\mathcal{O}^w = \mathcal{O}^-$.

Proposition 8.24. *Let* $\mathcal{O}^* = \{r^* | r \in \mathcal{O}\}$. *Then* $\mathcal{O} = \mathcal{O}^*$. *In particular,* $\mathcal{O}$ *is an inverse monoid.*

Proof. $r \in \mathcal{O}$ iff $rBr^* = rr^*Brr^*$ iff $r^*rBr^*r = r^*rr^*Brr^*r = r^*Br$ iff $r \in \mathcal{O}^*$. Hence $\mathcal{O} = \mathcal{O}^*$.

Lemma 8.25. *Let* $e\sigma \in R$. *Then there exists* $c \in C_W(e)$ *such that* $ce\sigma \in \mathcal{O}$. *In particular,* $ce\sigma \mathcal{H} e\sigma$.

Proof. $eB^\sigma e$ and eBe are Borel subgroups of H_e containing eT. Thus, there exists $c \in C_W(e)$ such that $c(eB^\sigma e)c^{-1} = eBe$. But $c(eB^\sigma e)c^{-1} = eB^{c\sigma}e$ and so $ce\sigma = ec\sigma \in \mathcal{O}$.

Proposition 8.26. *Let* $e, f \in E(R)$, $e\mathcal{J}f$. *Choose* $\sigma \in W$ *of minimal length such that* $\sigma^{-1}e\sigma = f$, *and let* $r = e\sigma$. *Then* $r \in \mathcal{O}$.

Proof. We proceed by induction on this minimal length l. If $l = 0$ then $e = f$ and the result follows from Lemma 8.21(c). So assume that $l = n+1$. We can write $e\sigma = e\tau\rho$ where $l(\tau) = n = l(\sigma)-1$. Now $e\tau \in \mathcal{O}$ since, if $\zeta^{-1}e\zeta = \tau^{-1}e\tau$ with $l(\zeta) < l(\tau)$, then $(\zeta\rho)^{-1}e(\zeta\rho) = \sigma^{-1}e\sigma$ with $l(\zeta\rho) < l(\sigma)$, a contradiction.

Also $\tau^{-1}e\tau\rho \neq \rho\tau^{-1}e\tau$ since otherwise $\tau^{-1}e\tau = \sigma^{-1}e\sigma$ which contradicts the minimality of $l(\sigma)$. So we show that $\tau^{-1}e\tau \cdot \rho \in \mathcal{O}$. Then by Lemma 8.21(a) $e\tau \cdot \tau^{-1}e\tau\rho = e\sigma \in \mathcal{O}$. Hence let $f = \tau^{-1}e\tau$. Then we have $f\rho \neq \rho f$. Now $\rho = \sigma_\alpha$ for some unique $\alpha \in \Delta$ and there exists a closed subgroup $V \subseteq B$ such that $B = VU_\alpha = U_\alpha V$ and $B^\rho = VU_{-\alpha} = U_{-\alpha}V$. By Lemma 5.1 of [105] we have that either $fU_\alpha = \{f\}$ and $U_{-\alpha}f = \{f\}$ or $U_\alpha f = \{f\}$ and $fU_{-\alpha} = \{f\}$. In either case $fBf = fVf = fB^\rho f$. Hence $f\rho \in \mathcal{O}$.

Theorem 8.27. *Suppose that $r, s \in R$, $r\mathcal{H}s$, and $r, s \in \mathcal{O}$. Then $r = s$.*

Proof. Let $r = e\sigma$ and $s = ce\sigma$ where $ce = ec$. Then $eB^\sigma e = eBe = eB^{c\sigma}e = c(eB^\sigma e)c^{-1}$. But then $ce \in N_{H_e}(eT) \cap eB^\sigma e$ and hence $ce = e$.

There are other ways to characterize the elements of $\mathcal{O}$.

Proposition 8.28. *Let $r = e\sigma = \sigma f \in R$. The following are equivalent:*

a) $r \in \mathcal{O}$,
b) $Br \cap rB = eBr$,
c) $Br \cap rB = rBf$,
d) $eBr = rBf$.

Proof. $r \in \mathcal{O}$ iff $eBr \subseteq rB$ iff $eBr \subseteq rB \cap Br$ (since always$eBr \subseteq Br$) iff $eBr = rB \cap Br$ (by an easy calculation). Similarily, (a) iff (c). But $Br \cap rB = eBr \cap rBf$ for any $r \in R$. Hence (b) iff (c).

Corollary 8.29. *Let $r \in \mathcal{O}$. Then $l(r) = \dim(Br) + \dim(rB) - 2\dim(eBr)$.*

Proof. Recall that $l(r) = \dim(BrB) - \dim(B\nu B)$. But $\dim(BrB) = \dim(Br) + \dim(rB) - \dim(Br \cap rB)$ while, by Proposition 8.28, $\dim(Br \cap rB) = \dim(B\nu B)$.

Remark 8.30. Write $r = e\sigma = \sigma f \in R$. Then

$$\begin{aligned} Br \cap rB &= eBr \cap rBf \\ &= eBe\sigma \cap \sigma fBf \\ &= (eBe \cap eB^\sigma e)\sigma. \end{aligned}$$

This pictures the elements of $\mathcal{O}$ as those for which $\dim(Br \cap rB)$ is maximal.

As usual we fix $B \subseteq G$ a Borel subgroup and $T \subseteq B$ a maximal torus. Recall from Proposition 8.15 that for each $W \times W$ orbit J of R there exists a unique $\nu \in J$ such that $B\nu = \nu B$. We now study the *Green's relations* on R and $\mathcal{O}$. Let $r, s \in R$.

Theorem 8.31. *Let $r \in R$. Then there exist unique elements $r_+, r_-, r_0 \in R$ such that*

a) $r = r_+ r_0 r_-$

b) $r_0 \mathcal{H} \nu$, *where* $\nu \mathcal{J} r$ *and* $B\nu = \nu B$
c) $r_+ \mathcal{R} r$ *and* $r_- \mathcal{L} r$
d) $r_+, r_- \in \mathcal{O}$.

Proof. First suppose that, for $r \in R$, we have $r = r_+ r_0 r_-$ satisfying a)-(d). If also $r = s_+ s_0 s_-$ then a), b) and c) imply that $r_+ \mathcal{H} s_+$ and $r_- \mathcal{H} s_-$. But then, by Theorem 8.27, $r_+ = s_+$ and $r_- = s_-$. Hence r_+ and r_- are unique. But then $r_0 = r_+^* r r_-^*$ and so r_0 is also unique.

To establish existence, we count. Let $\nu = e_0\sigma = \sigma f_0 \in J = WrW$ and let

$$A = \{x \in \mathcal{O} | x \mathcal{L} e_0\},$$

$$B = \{x \in \mathcal{O} | x \mathcal{R} f_0\}$$

and

$$C = \{x \in R | x \mathcal{H} \nu\}.$$

By the above, the product map

$$A \times C \times B \to WrW = W\nu W$$

is injective.

By Proposition 8.26 and Theorem 8.27 $|A| = |E(J)|$ and $|B| = |E(J)|$, while $|C| = |H_\nu|$, where $H_\nu = C$ is the $\mathcal{H}$-class of ν. Thus $|A \times C \times B| = |E(J)|^2 |H_\nu|$. To count up J we consider the map

$$\zeta : W\nu W \to E(J) \times E(J)$$

$\zeta(\sigma\nu\tau^{-1}) = (\sigma e_0 \sigma^{-1}, \tau f_0 \tau^{-1})$. It is easy to check that ζ is well defined and surjective, and all fibres have the same cardinality. But $\zeta^{-1}(e_0, f_0) = H_\nu$. Thus, $|J| = |E(J)|^2 |H_\nu|$.

8.6 The Adherence Order on *R*

Definition 8.32. *Let* $x, y \in R$. *We say that* $x \geq y$ *if* $\overline{BxB} \supseteq ByB$. *Clearly,* $(R, \leq)$ *is a poset.*

This is the obvious generalization of the Bruhat-Chevalley order from group theory to the case of reductive monoids. The main result of this section is the description of $(R, \leq)$ in terms of $(\Lambda, \leq)$ and $(W, \leq)$. Here,

$$e \leq f \text{ in } \Lambda \quad \text{if} \quad ef = fe = e,$$

and

$$x \leq y \text{ in } W \quad \text{if} \quad BxB \subseteq \overline{ByB}.$$

In section § 8.8 we go on to obtain a precise description of $(R, \leq)$ for $M = M_n(K)$.

So, back to work: for $e \in \Lambda$, let

$$
\begin{aligned}
W(e) &= C_W(e) \\
W_*(e) &= \{x \in W(e) \mid xe = e\} \lhd W(e) \\
D(e) &= \{x \in W \mid x \text{ has minimal length in } xW(e)\} \\
D_*(e) &\equiv \{x \in W \mid x \text{ has minimal length in } xW_e\}.
\end{aligned}
$$

Now if $v \in R$ we can write

$$v = xey^{-1}$$

where $x, y \in W$. But we can also assume that $y \in D(e)$. (by writing $y = y_0 w$, where $w \in W(e)$ and $y_0 \in D(e)$, so that $v = xey^{-1} = xe(y_0 w)^{-1} = xwey_0^{-1}$.)

The following result is recorded in detail, in [73]. We indicate here the main points of the proof.

Theorem 8.33. *Let $e, f \in \Lambda$, $x, s \in W$, $y \in D(e)$ and $t \in D(f)$. Then the following are equivalent.*

a) $xey^{-1} \le sft^{-1}$.
b) $ef = e$ and there exist $w \in W(f)W_*(e)$ and $z \in W_*(e)$ such that $tw \le y$ and $x \le swz$.

Proof. Assume that b) holds with w and z as indicated. Now $twey \in \overline{B}$ by (7) of [73]. Hence $sfwey^{-1} = sft^{-1}twey^{-1} \in sft^{-1}\overline{B}$. Thus $sfwey^{-1} \le sft^{-1}$. By assumption $w = w_1w_2$ with $w_1 \in W(f)$ and $w_2 \in W_*(e)$. Hence $sfwey^{-1} = sfw_1w_2ey^{-1} = sfw_1ey^{-1} = sw_1fey^{-1} = sw_1ey^{-1} = sw_1w_2ey^{-1} = swey^{-1}$. But from (6) of [73], $xey^{-1} \le swzey^{-1}$. Hence

$$xey^{-1} \le swzey^{-1} = swey^{-1} = sfwey^{-1} \le sft^{-1}\,.$$

Conversely, assume that $xey^{-1} \le sft^{-1}$. Clearly $e \le f$, and $xey^{-1} \in \overline{Bsft^{-1}B}$. Hence $e \in \overline{x^{-1}Bsft^{-1}By}$. Now there exists a unique $w \in W$ such that

$$A_w = B^-\widetilde{w}B \cap x^{-1}Bs \subseteq x^{-1}Bs$$

is open and dense. Thus,

$$e \in \overline{A_w ft^{-1}By}\,.$$

From here we can get $\widetilde{w}f\widetilde{w}^{-1}e = e$, and finally

$$\widetilde{w} = w_1w_2 \text{ where } w_1 \in W_*(e) \text{ and } w_2 \in W(f)\,.$$

But we are not not quite there. There exists a unique $u \in W$ such that

$$C_u = B^-uB \cap w_2t^{-1}By \subseteq w_2t^{-1}By$$

is open and dense.

It turns out that condition b) is satisfied with

$$w = w_2^{-1} u \in W(f)W_*(e) \;,$$

and

$$z = u^{-1} w_1^{-1} \in W_*(e) \;.$$

We can simplify the statement in the above theorem by using the following definition.

Definition 8.34. *Let $a \in R$. Then*

$$a = xey^{-1}$$

where $e \in \Lambda$, $x \in D_*(e)$ and $y \in D(e)$. We call this the normal form *of a.*

It easy easy to check that this expression for a is unique. For example, let $M = M_3(K)$. Then

$$\Lambda = \left\{ \begin{pmatrix} 1 & & \\ & 1 & \\ & & 1 \end{pmatrix}, \begin{pmatrix} 1 & & \\ & 1 & \\ & & 0 \end{pmatrix}, \begin{pmatrix} 1 & & \\ & 0 & \\ & & 0 \end{pmatrix}, \begin{pmatrix} 0 & & \\ & 0 & \\ & & 0 \end{pmatrix} \right\} = \{e_3, e_2, e_1, e_0\}$$

and

$$W = S_3 \;.$$

Let

$$a = \begin{pmatrix} 0\,1\,0 \\ 0\,0\,0 \\ 1\,0\,0 \end{pmatrix} .$$

Then

$$a = (132)e_2(1).$$

This is the normal form since

$$D_*(e_2) = W \qquad \text{and} \qquad D(e_2) = \{1, (23), (123)\} \;.$$

Corollary 8.35. *Let $a = xey^{-1}$ and $b = sft^{-1}$ be in normal form. Then the following are equivalent:*

a) $a \le b$
b) $e \le f$ and there exists $w \in W(f)W_*(e)$ such that $x \le sw$ and $tw \le y$.

Proof. If b) is satisfied then so too is Theorem 8.33 b) with $z = 1$. Conversely, given Theorem 8.33 b), we have $w \in W(f)W_*(e)$ and $z \in W_e$ with $tw \le y$ and $x \le swz$. But if $x_0 \in xW_*(e)$ is the element of minimal length then

$$x_0 \le sw\gamma \quad \text{for all} \quad \gamma \in W_*(e). \tag{1}$$

On the other hand, $W(f) = W^*(f) \times W_*(f)$ where $W^*(f) = \langle s \in S \mid sf = fs \rangle$, while $W_*(f) \subseteq W_*(e)$. Hence $w \in W(f)W_*(e)$ implies that $w = uv$ where $u \in W^*(f)$ and $v \in W_*(e)$. Let $s_0 \in xW_*(f)$ be the element of minimal length. Then

$$\begin{aligned} sw &= s_0\mu w \ , \quad \text{for some } \mu \in W_*(f) \\ &= s_0\mu uv \\ &= s_0 u\mu v \ , \quad \text{since } W_f \subseteq C_W(W^*(f)) \ . \end{aligned}$$

Hence $sw \in s_0 w W_*(e)$ since $\mu v \in W_*(e)$. Thus from (1) $x_0 \leq s_0 w$. Thus, b) is satisfied since $x_0 e y^{-1}$ (respectively, $s_0 f t^{-1}$) is the normal form for xey^{-1} (resp. sft^{-1}).

A special case of Corollary 8.35 has been obtained by Brion [12], Kato [46] and Springer [130], for the canonical monoid.

8.7 The j-order, R^+ and Pennell's Theorem

In this section we describe Pennell's theorem. This is a decomposition theorem for the adherence order on R. Unlike the situation of the Weyl group, we now have a principle of two types. The basic idea is easily described: we start with $x, y \in R$ and assume $x \leq y$. We then see that there exists $x_0, x_2, \ldots, x_n \in R$ such that $x = x_0 < x_2 <, \ldots, < x_n = y$ and $\ell(y) - \ell(x) = n$. Then $\ell(x_{i+1}) = \ell(x_i) + 1$. (That part is easy.) Pennell's idea is to establish the following dichotomy.

For each $x_i < x_{i+1}$ we have either

a) $x_i \in \overline{B} x_{i+1} \overline{B}$, or else
b) x_{i+1} is obtained from x_i via a "Bruhat interchange" (see Definition 8.41.

The key here is that, for each i, only one of a) and b) is true. The exact definition of b) will be given below. For $M_n(K)$ it implies that x_i and x_{i+1} have the same nonzero rows and columns. It is the natural generalization of the situation encountered in the Bruhat-Chevalley order on a Coxeter group [41]. The order relation of a) has no analogue in group theory. It was first studied by Pennell [72].

Definition 8.36. *a) Let $x, y \in R$. We write $x \leq_j y$ if $BxB \subseteq \overline{ByB}$. We refer to $\leq_j$ as the j-order on R.*
b) Define

$$\begin{aligned} R^+ &= \{x \in R \mid BxB \subseteq \overline{B}\} \\ &= \{x \in R \mid x \leq 1\} \ . \end{aligned}$$

Lemma 8.37. *Let $r = xey^{-1} \in R$ be in normal form. Then $r \in R^+$ if and only if $x \leq y$.*

Proof. If $r \in R^+$ then $r \leq 1$. Then by Corollary 8.35 there exists $w \in W$ such that $x \leq w \leq y$. Conversely, if $x \leq y$ then Corollary 8.35 b) is satisfied with $a = r$ and $b = 1$, by choosing $w = 1$.

We now find a discrete characterization of the j-order. Recall from Theorem 8.6 that for $x \in R$ and $s \in S$

$$sBr \subseteq BrB \cup BsrB$$

and

$$rBs \subseteq BrB \cup BrsB.$$

Lemma 8.38. *a) If $r \in R$ and $x \in W$ then $rBx \subseteq \bigcup_{\{y \in W | y \leq x\}} rBy$ and*

$$xBr \subseteq \bigcup_{\{y \in W | y \leq x\}} yBr \ .$$

b) If $r, r_1 \in R$ then $rBr_1 \subseteq \bigcup_{r_2 \leq r_1} Brr_2B$.

c) If $r, r_1 \in R$ then $r_1Br \subseteq \bigcup_{r_2 \leq r_1} Br_2rB$.

Proof. For a) we proceed by induction on $\ell(x)$. The case of $\ell(x) = 1$ follows from 8.6 since $S = \{x \in W \mid \ell(x) = 1\}$. So assume $x = s_1 s_2 \cdot \cdots \cdot s_k$ has length k. Then $rBx = (rBs_1 \cdot \cdots \cdot s_{k-1})s_k \subseteq (\cup_{y \leq x'} BryB)s_k$ where $x' = s_i \cdot \ldots s_{k-1}$. But $BryBs_k \subseteq BryB \cup Brys_kB$. Thus $rBx \subseteq \bigcup_{y \leq x'} (BryB \cup Brys_kB)$. The other case is similar.

For b) write $r_1 = xey^{-1}$ in normal form. Then by a) we have $rBr_1 = rBxey^{-1} \subseteq \cup_{x_1 \leq x} Brx_1Bey^{-1}$. But recall that $e \in \Lambda$, so that $B = eBe = C_B(e)e$. Hence

$$Brx_1Bey^{-1} = Brx_1C_B(e)ey^{-1} = Brx_1eC_B(e)y^{-1} \ .$$

But $y \in D(e)$ so that $yC_B(e)y^{-1} \subseteq B$. Thus

$$Brx_1C_B(e)e = Brx_1ey^{-1}yC_B(e)y^{-1} \subseteq Brx_1ey^{-1}B.$$

But recall that $x_1 \leq x$ and $y \in D(e)$, so that $r_2 = x_1ey^{-1} \leq xey^{-1} = r_1$. Finally, $rBr_1 \subseteq \cup_{x_1 \leq x} Brx_1Bey^{-1} \subseteq \cup_{x_1 \leq x} Brx_1ey^{-1}B \subseteq \cup_{r_2 \leq r_1} Brr_2B$.

For c) we can use the involution.

Theorem 8.39. *Let $x, y \in R$. Then $x \leq_j y$ if and only if $x \in R^+yR^+$.*

Proof. If $x \in R^+yR^+$ then $x = r_1yr_2$ for $r_1, r_2 \in R^+$. Thus $BxB = Br_1yr_2B \subseteq \overline{Br_1}\ y\ \overline{r_2B} \subseteq \overline{B}y\overline{B}$. Conversely, if $BxB \subseteq \overline{ByB}$, then there exist $r_1, r_2 \in R^+$ such that $BxB \subseteq Br_1ByBr_2B$. But it is easy, using Lemma 8.38 b) and c), to see that

$$Br_1ByBr_2B \subseteq \bigcup_{\substack{r_3 \leq r_1 \\ r_4 \leq r_2}} Br_3yr_4B \ .$$

We now describe where this dichotomy comes from. Define

$$Ref(W) = \bigcup_{w \in W} wSw^{-1} .$$

Lemma 8.40. *Let $\beta \in Ref(W)$, $x, y \in W$ and $e \in \Lambda$. Assume that $x < x\beta$. Then the following are equivalent.*

a) $Bxey^{-1}B \subseteq \overline{B}x\beta ey^{-1}\overline{B}$ and $xey^{-1} \neq x\beta ey^{-1}$
b) $\beta \notin W(e)$.

Proof. Assume b). If $\beta \notin W(e)$ then $\beta e \neq e$. Then $xey^{-1} \neq x\beta ey^{-1}$. After some calculation (as in Theorem 2.8 of [73]) we find that $xe\beta x^{-1} \in R^+$, using Lemma 8.37. Hence

$$\begin{aligned} Bxey^{-1}B &\subseteq \overline{B}xey^{-1}\overline{B} \\ &= \overline{B}xe\beta x^{-1}x\beta ey^{-1}\overline{B} \\ &\subseteq \overline{B}xe\beta x^{-1}\overline{B}x\beta ey^{-1}\overline{B} \\ &\subseteq \overline{B}x\beta ey^{-1}\overline{B} \end{aligned}$$

Conversely, assume that $\beta \in W(e)$. Then one obtains after some elementary calculation that either $xey^{-1} = x\beta ey^{-1}$ or else $Bxey^{-1} \not\subseteq \overline{B}x\beta ey^{-1}\overline{B}$.

Definition 8.41. *Let xey^{-1}, set^{-1} be in normal form with $xey^{-1} \neq set^{-1}$. We say set^{-1} is obtained from xey^{-1} via a* Bruhat interchange *if there exists $\beta \in Ref(W) \cap W(e)$ such that $x \leq x\beta$ and $x\beta ey^{-1} = set^{-1}$.*

We see from Lemma 8.40 that this is the case where $xey^{-1} \not\leq_j set^{-1}$.

Example 8.42. In $M_3(K)$ consider

$$xcy^{-1} = \begin{pmatrix} 1&0&0 \\ 0&1&0 \\ 0&0&1 \end{pmatrix} \begin{pmatrix} 1&0&0 \\ 0&1&0 \\ 0&0&0 \end{pmatrix} \begin{pmatrix} 1&0&0 \\ 0&0&1 \\ 0&1&0 \end{pmatrix} = \begin{pmatrix} 1&0&0 \\ 0&0&1 \\ 0&0&0 \end{pmatrix}$$

and

$$x\beta ey^{-1} = \begin{pmatrix} 1&0&0 \\ 0&1&0 \\ 0&0&1 \end{pmatrix} \begin{pmatrix} 0&1&0 \\ 1&0&0 \\ 0&0&1 \end{pmatrix} \begin{pmatrix} 1&0&0 \\ 0&1&0 \\ 0&0&0 \end{pmatrix} \begin{pmatrix} 1&0&0 \\ 0&0&1 \\ 0&1&0 \end{pmatrix} .$$

Theorem 8.43. (Pennell's Theorem) *Let $a, b \in R$. Then $a \leq b$ if and only if there exist $\theta_0, \theta_1, \ldots, \theta_r \in R$ such that $a = \theta_0 \leq \theta_1 \leq \cdots \leq \theta_r = b$ and, for each ℓ, either $\theta_{\ell-1} \leq_j \theta_\ell$ or else θ_ℓ is obtained from $\theta_{\ell-1}$ by a Bruhat interchange.*

Proof. Plainly, the condition is sufficient. Assume therefore that $a \leq b$. Write $a = xey^{-1}$ and $b = sft^{-1}$ in normal form. Then from Theorem 8.43 there exists $w = w_1w_2 \in W(f)W_e$ such that $x \leq sw$ and $tw \leq y$. Now

$$\begin{aligned} sfwey^{-1} &= sfw_1w_2ey^{-1} \\ &= sw_1few_2y^{-1} \\ &= sw_1ew_2y^{-1} \\ &= swey^{-1} . \end{aligned}$$

Hence

$$\begin{aligned} Bswey^{-1} &\subseteq \overline{B}sft^{-1}twey^{-1}\overline{B} \\ &\subseteq \overline{B}sft^{-1}\overline{B} \end{aligned}$$

since $twey^{-1} \in R^+$.

But also $xey^{-1} \leq swey^{-1}$ since $x \leq sw$. We therefore obtain

$$xey^{-1} \leq swey^{-1} \leq_j \ sft^{-1} ,$$

and we are reduced to finding the chain of θ_i's from xey^{-1} to $swey^{-1}$. Since $x \leq sw$ we can find (as in Proposition 5.11 of [41]) $\gamma_1, \dots, \gamma_r \in Ref(W)$ such that

$$x < x\gamma_1 < x\gamma_1\gamma_2 < \cdots < x\gamma_1 \cdot \cdots \cdot \gamma_r = sw$$

and, for each $i = 1, \dots, r$, there exists $\delta_i \in W_e$ such that $x\gamma_1 \cdot \cdots \cdot \gamma_i\delta_i \in D_e$. Since $x \in D_e$ we obtain

$$x \leq x\gamma_1\delta_1 \leq x\gamma_1\gamma_2\delta_2 \leq \cdots \leq x\gamma_1 \cdot \cdots \cdot \gamma_r\delta_r .$$

If $x = x\gamma_1\gamma_2 \cdot \cdots \cdot \gamma_r\delta_r$ then

$$swey^{-1} = sw\delta_r^{-1}ey^{-1} = x\gamma_1 \cdot \cdots \cdot \gamma_r ey^{-1} = xey^{-1}$$

and so we are done.

If $x \neq x\gamma_1 \cdot \cdots \cdot \gamma_r\delta_r$ then let

$$\theta_i = x\gamma_1 \cdot \cdots \cdot \gamma_i\delta_i ey^{-1} .$$

If $\gamma_i \in Ref(W)\backslash W(e)$ then by Lemma 8.40 $B\theta_{i-1}B \subseteq \overline{B}\theta_i\overline{B}$ while, if $\gamma_i \in Ref(W) \cap W(e)$, θ_i is obtained from θ_{i-1} via a Bruhat interchange.

Example 8.44. With $M = M_4(K)$ let

$$a = \begin{pmatrix} 0&0&1&0 \\ 0&0&0&1 \\ 0&0&0&0 \\ 0&0&0&0 \end{pmatrix} \leq \begin{pmatrix} 0&0&0&1 \\ 0&0&0&0 \\ 0&1&0&0 \\ 0&0&0&0 \end{pmatrix} = b .$$

Then

$$\theta_0 = \begin{pmatrix} 0&0&1&0 \\ 0&0&0&1 \\ 0&0&0&0 \\ 0&0&0&0 \end{pmatrix} \leq \theta_1 = \begin{pmatrix} 0&1&0&0 \\ 0&0&0&1 \\ 0&0&0&0 \\ 0&0&0&0 \end{pmatrix} \leq \theta_2 = \begin{pmatrix} 0&1&0&0 \\ 0&0&0&0 \\ 0&0&0&1 \\ 0&0&0&0 \end{pmatrix} \leq \theta_3 = \begin{pmatrix} 0&0&0&1 \\ 0&0&0&0 \\ 0&1&0&0 \\ 0&0&0&0 \end{pmatrix} .$$

One can easily see that

$$\theta_0 \leq_j \theta_1 \leq_j \theta_2 ,$$

while $\theta_2 \leq \theta_3$ via a Bruhat interchange. The reader is encouraged to observe how each θ_i is obtained from θ_{i-1} by way of an "elementary" move. This example is developed in complete detail in the next section.

8.8 The Adherence Order on $M_n(K)$

In this section we use Pennell's Theorem to determine $(R, \leq)$ for $M = M_n(K)$. The calculation is straightforward. We simply calculate $\leq_j$ combinatorially, and then characterize the Bruhat interchange for $R_n = R(M_n(K))$.

Let E_{ij} be the $n \times n$ matrix (a_{st}) such that

$$a_{st} = \begin{cases} 1 \,, \ (s,t) = (i,j) \\ 0 \,, \ (s,t) \neq (i,j) \end{cases} .$$

We refer to E_{ij} as an *elementary matrix.*

Proposition 8.45. *a) $E_{ij} \leq E_{km}$ if and only if $i \leq k$ and $m \leq j$.*
b) Let $A, C \in R_n$ and write

$$A = \sum_{\ell=1}^{s} A_\ell$$

and

$$C = \sum_{\ell=1}^{t} C_\ell$$

where $\{A_\ell\}$ and $\{C_\ell\}$ are elementary matrices. Define $S_A = \{A_1, \dots, A_s\}$ and $S_C = \{C_1, \dots, C_t\}$. Then the following are equivalent:
i) $A \leq_j C$
ii) There exists an injection $\theta : S_A \longrightarrow S_C$ such that $A_\ell \leq_j \theta(A_\ell)$ for each $\ell = 1, \dots, s$.

Proof. If $E_{ij} \leq E_{km}$ then there exist upper triangular matrices X and Y such that $XE_{km}Y = E_{ij}$. But this implies $(XE_{km}Y)_{st} = 0$ if $s > k$ or $t < m$. Conversely, if $i \leq k$ and $m \leq j$ then $X = E_{ik}$ and $Y = E_{mj}$ are both upper triangular. But then $XE_{k\ell}Y = E_{ij}$.

For b) first assume that $A \leq_j C$. Notice that $s = rank(A)$ and $t = rank(C)$. Write

$$A = w_1 \begin{pmatrix} I_s & 0 \\ 0 & 0 \end{pmatrix} w_2$$

and

$$C = w_3 \begin{pmatrix} I_t & 0 \\ 0 & 0 \end{pmatrix} w_4 \,,$$

where $w_1, w_2, w_3, w_4 \in S_n$, the unit group of R_n. Let $x = w_1 w_3^{-1}$ and $Y = w_4^{-1} \begin{pmatrix} I_s & 0 \\ 0 & 0 \end{pmatrix} w_2$. Then $XCY = A$, and one checks that, for each $k = 1, \ldots, t$,

$$XC_kY \leq_j C_k$$

and

$$XC_kY \in S_A \cup \{0\}.$$

Then define $\theta(A_\ell) = C_k$ such that $XC_kY = A_\ell$.

Conversely, given an injection $\theta : S_A \longrightarrow S_C$ such that $A_\ell \leq_j \theta(A_\ell)$ for $i = 1, \ldots, S$, then from a) there exist elementary upper triangular matrices X_ℓ and Y_ℓ such that $X_\ell \theta(A_\ell) Y_\ell = A_\ell$. Let $X = \sum_{\ell=1}^{s} A_\ell$ and $Y = \sum_{\ell=1}^{s} Y_\ell$. Then $XCY = A$.

Proposition 8.46. *Let $A = E_{i_1j_1} + \cdots + E_{i_sj_s} \in R_n$ and suppose*

$$C = E_{i_1j_1} + \cdots + E_{i_{h-1}i_{h-1}} + E_{i_\ell j_\ell} + E_{i_{h+1}j_{h+1}} + \cdots + E_{i_{\ell-1}j_{\ell-1}} \\ + E_{i_hj_h} + E_{i_{\ell+1}j_{\ell+1}} + \cdots + E_{i_sj_s}$$

so that C is obtained from A by interchanging two nonzero rows. Relabel the two altered positions of A by E_{ij} and $E_{k\ell}$. Then $A < C$ if and only if $i < k$ and $j < \ell$ or $i > k$ and $j > \ell$.

Proof. First assume that $s = 2$. Then

$$A = (\ell-1\; i) \cdot \cdots \cdot (12)(k-1\; h) \cdot \cdots \cdot (34)(23)E(23) \ldots (\ell-1\; \ell)(12) \cdot \cdots \cdot (j-1\; j)\,.$$

This is the normal form for A. One checks that

$$C = [(i-1\; i) \cdot \cdots \cdot (12)(k-1\; h) \cdot \cdots \cdot (34)(23)(12)] \\ \ldots E(23) \cdot \cdots \cdot (\ell-1\; \ell)(12) \cdot \cdots \cdot (j-1\; j)\,.$$

But $(i-1\; i) \cdot \cdots \cdot (23) < (i-1\; i) \cdot \cdots \cdot (23)(12)$ and so $A < C$. Conversely, if $i < k$ and $j > \ell$ or $i > k$ and $j < \ell$ then the same argument shows that $C < A$.

The proof of the general case ($s > 2$) is similar.

Proposition 8.46 describes the situation of a Bruhat interchange for $M_n(K)$. It says that $A < C$ via a Bruhat interchange if and only if

a) C is obtained from A by interchanging two non-zero rows of A.
b) In the process of a), a 2×2 submatrix $\begin{pmatrix} 1 & 0 \\ 0 & 1 \end{pmatrix}$ of A ends up as $\begin{pmatrix} 0 & 1 \\ 1 & 0 \end{pmatrix}$.

Theorem 8.47. *Let $x, y \in R_n$ and assume that $x \leq y$ with $\ell(x) = \ell(y) - 1$. Then one of the following holds.*

a) x is obtained from y by setting some nonzero entry of y to zero.
b) x is obtained from y by moving a nonzero entry either upward or to the right.
c) y is obtained from x via a Bruhat interchange.

Proof. Assume that c) is not the case and $rank(x) < rank(y)$. Then by Theorem 8.43 and Proposition 8.45 there is an injection $\theta : S_x \longrightarrow S_y$ such that $x_i \leq \theta(x_i)$ for each $x_i \in S_x$. So define $\theta(x) \in R_n$ by $S_{\theta(x)} = \theta(S_x)$. Then $x \leq \theta(x) < y$. Thus $x = \theta(x)$, and so a) holds.

Now assume that c) is not the case, and $rank(x) = rank(y)$. By Pennell's Theorem $x \leq_j y$, and so there exist $a, b \in R^+$ such that $x = ayb$. Then $x = ayb \leq yb \leq b$. Hence, either $x = yb$ or else $yb = b$. In the first case $x = yb$, and in the second case $x = ay$. Assume without loss of generality that $x = yb$ (x and y have the same nonzero rows). An elementary argument shows that x is obtained from y by moving some nonzero entry of y to the right. Thus b) holds.

Theorem 8.47 allows us to give a combinatorial description of the adherence order on R_n. First we represent the elements of R_n by sequences of non-negative integers. Given $x \in R_n$ we associate with x a sequence $(\epsilon_1, \epsilon_2, \ldots, \epsilon_n)$ where for each i, $1 \leq i \leq n$,

$$\epsilon_i = \begin{cases} 0 \text{ if } x \text{ has all zeros in the } i^{th} \text{ column} \\ j \text{ if } x_{ji} = 1 \,. \end{cases}$$

For example, $\begin{pmatrix} 0\,0\,0\,0 \\ 0\,0\,0\,1 \\ 1\,0\,0\,0 \\ 0\,0\,1\,0 \end{pmatrix}$ corresponds to (3042). Notice that a sequence $(\epsilon_1, \epsilon_2, \ldots, \epsilon_n)$ occurs this way for some $x \in R_n$ if and only if $0 \leq \epsilon_i \leq n$ for each i; and whenever $\epsilon_i = \epsilon_j$, either $i = j$ or else $\epsilon_i = \epsilon_j = 0$.

Theorem 8.48. *Let $x = (\delta_1, \ldots, \delta_n)$, $y = (\epsilon_1, \ldots, \epsilon_n) \in R_n$. Then $\leq$ is the smallest partial order on R_n generated by declaring $x < y$ if either*

a) $\delta_j = \epsilon_j$ for $j \neq i$ and $\delta_i < \epsilon_i$, or
b) i) $\delta_k = \epsilon_k$ if $k \notin \{i, j\}$
ii) $i < j$
iii) $\delta_i = \epsilon_j$, $\delta_j = \epsilon_i$ and $\epsilon_i > \epsilon_j$.

From Theorem 8.47 we have three possibilities. We indicate how each one fits into the different possibilities of Theorem 8.48.

Proof. Theorem 8.47 a) falls under part a), since we can take $\delta_i = 0$.

Theorem 8.47 b) falls under part b) if x and y have the same nonzero rows; and under part a) if they have the same nonzero columns.

Theorem 8.47 c) is the special case of part b) where both ϵ_i and ϵ_j are nonzero.

Example 8.49. Let $x = (21403)$ and $y = (35201)$ in R_5. Then $x < y$ since

$$\begin{aligned}(21403) &< (31402)\ , \text{ by } 8.48b) \\ &< (34102)\ , \text{ by } 8.48b) \\ &< (35102)\ , \text{ by } 8.48a) \\ &< (35201)\ , \text{ by } 8.48b)\ .\end{aligned}$$

8.9 Exercises

1. Give an explicit description, for $M_n(K)$, of each of the following objects: $\Lambda, R, E(\overline{T}), \Phi, B, T, P_I$.
2. Let M be reductive with zero, and let $x \in M, x \neq 0$. Prove there exist $e, f \in E_1(\overline{T})$ such that $exf \neq 0$.
3. Let M be reductive. Prove that

$$M = \bigsqcup_{e \in E(\overline{T})} GeB.$$

4. For the inverse semigroup S recall the *natural order* $\leq$ on S: we say that $x \leq y$ if there exist $e, f \in E(S)$ such that $ey = yf = x$. Let M be reductive, and let $N = \overline{N_G(T)}$. Prove that $x \leq y$ in N, for the natural order, if and only if $x \in y\overline{T}$.
5. Let R^+ be the image of $\overline{B} \cap \overline{N_G(T)}$ in R. Prove that R^+ is $\mathcal{J}$-trivial. i.e. for $x, y \in R^+$, $x\mathcal{J}y$ if and only if $x = y$.
6. Assume that M is reductive, and let $\lambda : K^* \to M$ be a morphism. Define $w : M \to Aut(M)$ by $w(t)(x) = txt^{-1}$. Let $M_0 = \{x \in M | w(t)(x) = x\}$ and let
$M_+ = \{x \in M \mid lim_{t \to 0} w(t)(x) \ exists \ in \ M\}$.
   a) Find an example where M_0 is not irreducible.
   b) Prove that M_+ contains a Borel subgroup of M.
   c) Prove that M_0 and M_+ are submonoids of M.
   d) Prove that $\psi : M_+ \to M_0$, $\psi(x) = lim_{t \to 0} w(t)(x)$ is a surjective morphism of algebraic monoids.
   e) Let $B_0 \subseteq C_G(\lambda)$ be a Borel subgroup. Prove that $B_0 \backslash M_0 / B_0$ is finite. Can you identify the orbits?
7. Let M be semisimple, and let $H \subseteq G$ be a torus. Let $Z \subseteq G$ be the connected center of G. Let $M^* = \{x \in M \mid xH \subseteq Zx\ \}$. Notice that M^* is closed in M.

a) Let $e \in E(M^*) = E(M) \cap M^*$. Prove that there exists $f \in E(M^*)$, $f\mathcal{L}e$, such that $sf = fs$ for all $s \in H$. Hint: $H \subseteq C^l_G(e)$.

b) Let $\Lambda \subseteq E(\overline{T}) \subseteq \overline{C_G(H)}$ be a cross section lattice. Prove that $M^* = \bigcup_{f \in \Lambda^*} GfC_G(H)$, where $\Lambda^* = \Lambda \cap M^*$. Hint: use Exercise 6 above.

8. Let M be a reductive monoid with zero and assume that M has the property that

$$\Lambda(B,T) =: \{e \in E(\overline{T}) \mid Be \subseteq eB\} = \{f \in E(\overline{T}) \mid f \geq e_0\} \cup \{0\}$$

for some $e_0 \in E_1(\overline{T})$. Prove that R has the following property: for any $x \in R$, either $x^2 = 0$ or else $x\mathcal{H}e$ for some idempotent $e \in E(R)$.

9. Let M be reductive, and let $x \in M$ be an element of some minimal $\mathcal{J}$-class of M. Prove that, for some maximal torus T of M, $xT = Tx$.

10. Let M be reductive, and let $x \in M$ be such that $xT = Tx$ for some maximal torus T of M. Prove that $x \in \overline{N_G(T)}$.

11. Let $M = M_n(K)$. Prove that R has the following property: if $e \leq f < 1$ are idempotents of R, then there exists $r \in R$ such that $r^s = e$ for some $s > 0$, and r is $\mathcal{R}$-related to f.

12. Let G be a connected, solvable group acting on an irreducible, affine variety X. Suppose that there are a finite number of orbits. Let $\mathcal{O}$ be the set of G-orbits on X. Define $x \leq y$, if $x \subseteq \overline{y}$. Prove that $\mathcal{O}$ is a ranked poset; i.e. any two maximal chains from x to y have the same length. This applies, in particular, to the Bruhat decomposition of a reductive monoid.

13. Let $e \in E(\overline{T})$, $T \in B$. Prove that $BeB \cap H_e = C_B(e)e$, where H_e is the $\mathcal{H}$-class of e in M.

14. Let M be irreducible with unit group G, $e \in E(\overline{T})$, $T \in B$. Let U be the unipotent radical of $C_B(e)$. Define $U_1 = < U_\alpha \mid eU_\alpha = U_\alpha e \neq \{e\} >$ and $U_2 = < U_\alpha \mid eU_\alpha = U_\alpha e = \{e\} >$. Define $\psi : U_1 \times U_2 \to U$, by $\psi(a,b) = ab$. Prove that ψ is an isomorphism of groups.

15. Let M be reductive, and let $B \subseteq G$ be a Borel subgroup of G. Let J be a regular $\mathcal{J}$-class of $\overline{B}$. Define $J^* = \{a \in M \mid a\mathcal{H}x \text{ for some } x \in J\}$. Prove that $J^* = BH_eB$, where H_e is the $\mathcal{H}$-class of e in M.

16. Let M be reductive and let $e \in E(\overline{T})$. Prove that

$$dim(GeG) = |\Phi| - |\{\alpha \in \Phi \mid s_\alpha e = es_\alpha = e\}| + dim(eT).$$

Hint: $BeB^- \subseteq GeG$ is open for appropriate, opposite Borel subgroups containing T. So compute $\dim(BeB^-)$.

17. Let M be reductive with zero. Define rank(x)=dim(xT) for $xT \in R$. For $e \in E_1(\overline{T})$, define $\chi_e : \overline{T} \to e\overline{T} \cong K$ by $\chi_e(z) = ez$. Assume that M has a unique minimal nonzero $\mathcal{J}$-class. Define a bijection between the rank-one elements of R and $\{s(\chi_e) - \chi_e \mid e \in E_1(\overline{T}),\ s \in W\} \backslash \{0\}$.

9

Representations and Blocks of Algebraic Monoids

The ultimate mathematical dictionary might define the theory of algebraic monoids as *a branch of algebra that determines the content of mathematical problems relating convexity and positivity to representation theory.* We can regard this imaginary definition as the major theme of this chapter. What I want to do here is provide the reader with a self-contained overview of what is known about algebraic monoids and their finite dimensional representations. In the first section we discuss normal, reductive monoids. Here we find that there is a perfect analogue of many results about reductive groups. In particular, we obtain the desired relationship between the set of irreducible representations of M and the adjoint quotient of M.

In the next section we focus on the special properties of a normal, reductive monoid M in characteristic $p > 0$. The results here are largely due to S. Doty. We find that $K[M]$ has a *good* filtration in the sense of Donkin [24]. Furthermore, the category of rational M-monoids is a *highest weight category* in the sense of Cline, Parshall and Scott [16].

In the last two sections we study the *blocks* of an algebraic monoid. Blocks are often better behaved for monoids than they are for groups. We discuss the blocks of monoids in two contrasting situations; solvable monoids with zero, and $M = M_n(K)$ when $char(K) = p > 0$. The blocks of a solvable monoid were studied by the author in [113]. The blocks of $M_n(K)$ were calculated by S. Donkin in [25].

9.1 Conjugacy Classes and Adjoint Quotient

Semisimple elements play an important rôle in the representation theory of reductive groups. The most fundamental results relate the conjugacy classes of semisimple elements to the characters of irreducible representations, via the ring of class functions on the adjoint quotient. In this section we discuss the analgous results for normal reductive monoids.

Let M be an irreducible normal, algebraic monoid. An element $x \in M$ is *semisimple* if $\rho(x) \in M_n(K)$ is diagonalizable for any rational representation $\rho : M \longrightarrow M_n(K)$ of M.

Proposition 9.1. *Let M be irreducible. The following are equivalent.*

a) $x \in M$ is semisimple.
b) $x \in \overline{T}$ for some maximal torus $T \subseteq G$.

Proof. b) clearly implies a). So assume a). Assume that $M \subseteq M_n(K)$ as a closed submonoid. Then $x \in M_n(K)$ is diagonalizable, and it follows easily that $x \in H_e$, the unit group of eMe, for some idempotent $e \in M$. But x is semisimple in this reductive group and so $x \in S$ where $S \subseteq H_e$ is a maximal torus. However, any maximal torus S of H_e is of the form $S = eT$ for some maximal torus $T \subseteq C_G(e)$. But $eT \subseteq \overline{T}$. Then $x \in S = eT \subseteq \overline{T}$.

Theorem 9.2. *Suppose that M is reductive and $x \in M$. Then the following are equivalent.*

a) x is semisimple.
b) $x \in \overline{T}$ for some maximal torus T of G.
c) $C\ell(x)$, the conjugacy class of x, is closed in M.

Proof. Suppose $x \in \overline{T}$. Then $txt^{-1} = x$ for $t \in T$. Thus, by 2.13, Corollary 1 of [134], $C\ell(x) \subseteq M$ is closed.

Conversely, assume that $C\ell(x) \subseteq M$ is closed. We assume also that M has a zero element. The general case follows easily from this. By [62] the categorical quotient $\pi : M \longrightarrow X$ of M by $G \times M \longrightarrow M$, $(g, x) \longmapsto gxg^{-1}$, exists and induces a one-to-one correspondence between the closed orbits of M and the points of X. Consider

$$\pi \mid \overline{T} \, : \, \overline{T} \longrightarrow X \, .$$

If $\pi(z) = \pi(y)$ then $gyg^{-1} = z$ for some g. But $g^{-1}Tg$ and T are both contained in $C_G(y)^0$. Hence there exists $h \in C_G(y)^0$ such that $hg^{-1}Tgh^{-1} = T$. But then $hg^{-1} \in N_G(T)$ while $(hg^{-1})^{-1}g\ hg^{-1} = gyg^{-1} = z$. Thus $\pi|\overline{T}$ induces an injective map

$$\theta \, : \, \overline{T}/W \longrightarrow X$$

where $W = N_G(T)/T$. On the other hand θ is known to be birational by Corollary 2 of [134]. Hence θ is an open embedding, topologically. Since M has a zero, $\overline{T}/W$ and X are cones. It follows that θ is a finite bijective morphism. If M is normal then so is X, and thus θ is an isomorphism by Theorem 2.29.

Recall now the element $x \in M$ with $C\ell(x) \subseteq M$ closed. By the above, $\pi(x) = \pi(y)$ for some $y \in \overline{T}$. But π separates closed orbits and so $C\ell(x) = C\ell(y)$.

Corollary 9.3. *Suppose that M is reductive and let $T \subseteq G$ be a maximal torus. Then*

$$\begin{aligned} C_M(T) &= \{x \in M \mid xt = tx \text{ for all } t \in T\} \\ &= \overline{T}\,. \end{aligned}$$

Proof. If $x \in C_M(T)$ then $C\ell(x) \subseteq M$ is closed and so, by Theorem 9.2, x is semisimple. It is then possible to find a Borel subgroup $B \subseteq G$ such that $T \cup \{x\} \subseteq \overline{B}$. From there we embed $\overline{B}$ in $T_n(K)$ as a closed submonoid. It follows from Proposition 15.4 [40] that $T \cup \{x\}$ is contained in a maximal torus of $T_n(K)$. Hence $T \cup \{x\} \subseteq T_n(K) \underset{\pi}{\longrightarrow} D_n(K)$ is injective, where π is the projection to the diagonal. But from Theorem 6.1, $\pi|\overline{B}$ factors through the universal morphism of $\overline{B}$ to a D-monoid $\theta : \overline{B} \longrightarrow \overline{T}$ and thus $x \in \overline{T}$.

Corollary 9.4. *Let $\pi : M \longrightarrow X$ be the categorical quotient induced by conjugation of G on M. Then in the following diagram θ is an isomorphism:*

$$\begin{array}{ccc} \overline{T} & \hookrightarrow & M \\ \big\downarrow & & \big\downarrow \pi \\ \overline{T}/W & \underset{\theta}{\longrightarrow} & X\,. \end{array}$$

Proof. This is included in the proof of Theorem 9.2.

Example 9.5. Let M be a normal monoid with 0 and unit groups $G\ell_2(K)$. By [103], $M \cong M_r$, $r \in \mathbb{Q}^+$, where M_r is the unique semisimple monoid with

$$X(\overline{T}) \cong \{(a,b) \in \mathbb{Z}^2 \;\Big|\; \left|\frac{b-a}{b+a}\right| \leq r\} \cup \{(0,0)\}\,.$$

Assume further that $r = 1/n$ with $(2,n) = 1$. Then

$$K[\overline{T}] \cong K[x,u,v]/(x^n - uv)$$

and the non-trivial element σ of the Weyl group $W = \{1, \sigma\}$ acts by

$$\begin{aligned} \sigma(x) &= x, \\ \sigma(u) &= v \end{aligned}$$

and

$$\sigma(v) = u.$$

Hence, the ring of invariants is

$$K[X] \cong K[x,u,v]^W = K[x,\ u+v]\,.$$

Thus

$$\pi \ :\ M_r \longrightarrow X \cong K^2$$

is a flat morphism.

The irreducible representations of a normal monoid can be calculated using Theorem 9.2. Let M be normal and reductive. We obtain

$$X(\overline{T}) \subseteq X(T), \text{ the set of characters of } \overline{T},$$

and

$$X(\overline{T})_+ \subseteq X(\overline{T}), \text{the set of dominant weights of } \overline{T}.$$

$X(\overline{T})_+$ is obtained by intersecting the set $X(T)_+$ of dominant weights of $X(T)$ with $X(\overline{T})$. As expected, $X(\overline{T})_+$ is a fundamental domain for the action of W on $X(\overline{T})$.

Theorem 9.6. *Let M be reductive and normal. Then there is a canonical one-to-one correspondence between $X(\overline{T})_+$ and the set of irreducible representations of M.*

Proof. Assume that $\rho : M \longrightarrow End(V)$ is irreducible. Then $\rho|G$ is irreducible since $G \subseteq M$ is dense. Thus $\rho|G$ is identified by its highest weight $\lambda \in X(T)_+$. But $\rho|G$ came from ρ, and so $\lambda \in X(\overline{T})$. Hence $\lambda \in X(\overline{T})_+$. Conversely, given $\lambda \in X(\overline{T})_+$ we start with $\rho_\lambda : G \longrightarrow G\ell(V_\lambda)$, the unique irreducible representation with highest weight λ. But all the other weights of ρ_λ are in the convex hull of $W \cdot \lambda$, which is contained in $X(\overline{T})$. Hence $\rho_\lambda|T$ extends over $\overline{T}$. Thus, by Theorem 5.2, ρ_λ extends to $\rho_\lambda : M \longrightarrow End(V_\lambda)$.

The reader who wants more detailed information about rational representations should consult [26, 107].

9.2 Rep(M) according to Doty

In this section we discuss some results of S. Doty [26] concerning the structure of $K[M]$ as a $G \times G$-module. This is not much of an issue if $char(K) = 0$ since, in that case, $K[M] = \bigoplus_{\lambda \in L(M)} K[M]_\lambda$, and each $K[M]_\lambda$ is $G \times G$-irreducible. Furthermore, $L(M)$ is canonically identified with the set of high weights of G that come from representations of M. If M is normal, we see that $L(M) = X(\overline{T})_+$, as in Theorem 9.6.

On the other hand, if $char(K) = p > 0$, then the situation is more complex. First of all, it is no longer sufficient to consider just the simple $G \times G$-modules. This leads us naturally to the theory of *highest weight categories*. From there we can better understand $K[M]$ in terms of filtrations.

Let $\lambda : B \longrightarrow K^*$ be a character of the Borel subgroup B of G. Define

$$H^0(\lambda) = ind_B^G(K_\lambda) = \left\{ f \in K[G] \middle| \begin{array}{l} f(bg) = \lambda(b)f(g) \text{ for all} \\ b \in B,\, y \in G \end{array} \right\}.$$

$H^0(\lambda)$ is naturally a G-module via $g \cdot f(x) = f(xg)$ for $x, g \in G$. It is well known, from Borel-Weil theory (see Remark 2.49), that $H^0(\lambda)$ is nonzero if and only if λ is dominant. In this case, the unique, maximal completely reducible submodule of $H^0(\lambda)$ is

$$V_\lambda \subseteq H^0(\lambda),$$

the irreducible G-module with highest weight λ. V_λ is the *socle* of $H^0(\lambda)$.

Now let M be normal and reductive. Define

$$ind_{\overline{B}}^M(K_\lambda) = \left\{ f \in K[M] \,\middle|\, \begin{array}{l} f(xy) = \lambda(x)f(y) \text{ for all} \\ x \in \overline{B},\ y \in M \end{array} \right\}.$$

It follows from Theorem 9.6 that $ind_{\overline{B}}^M(K_\lambda) \neq (0)$ if and only if $\lambda \in X(\overline{T})_+$, the set of dominant weights of $X(\overline{T})$.

Definition 9.7. *a) A* good filtration $(0) = V_0 \subseteq V_1 \subseteq V_2 \subseteq \ldots$ *of the rational G-module $V = \bigcup_{i\geq 0} V_i$ is one for which $V_{i+1}/V_i \cong H^0(\lambda_i)$ for each i, where λ_i is a dominant weight.*

b) Let $X(T)_+$ be the set of dominant weights of T. If $\pi \subseteq X(T)_+$ and V is a rational G-module, we say V belongs to π *if any G-module V_λ of the Jordan-Hölder series of V has $\lambda \in \pi$. In any case we let*

$$O_\pi(V) \subseteq V$$

be the maximal submodule of V belonging to π.

Proposition 9.8. *Let $\pi = X(\overline{T})_+$.*

a) $O_\pi H^0(\lambda) = \begin{cases} H^0(\lambda) & \text{if } \lambda \in \pi \\ (0) & \text{if } \lambda \notin \pi \,. \end{cases}$

b) Let $I(\lambda)$ (resp. $Q(\lambda)$) be the injective hull of V_λ in the category of rational G-modules (resp. M-modules). Then

$$O_\pi(I(\lambda)) = \begin{cases} Q(\lambda) \,, & \lambda \in \pi \\ (0) \,, & \lambda \notin \pi \,. \end{cases}$$

Furthermore, $Q(\lambda)$ has a good filtration with all factors V_μ having $\mu \geq \lambda$.

c) $O_\pi K[G] = K[M]$. Furthermore, $K[M]$ has a good filtration as a left G-module. Each $H^0(\mu)$, $\mu \in \pi$, occurs exactly dim $H^0(\mu)$ times.

Proof. For a), notice that any submodule of $H^0(\lambda)$ contains V_λ, and so, if $\lambda \notin \pi$, then $O_\pi H^0(\lambda) = (0)$. Now suppose that $\lambda \in \pi$. Then $H^0(\lambda)$ lifts (by Theorem 5.2) to become an M-module. Hence $O_\pi H^0(\lambda) = H^0(\lambda)$.

For b), first notice that $O_\pi(-)$ takes injectives to injectives by (1.1d) of [24]. Then the formula for $O_\pi(I(\lambda))$ follows, and $Q(\lambda)$ has a good filtration by Theorem 8 of [24].

c) is proved by "Frobenius" reciprocity. See Theorem 4.4 of [26].

Corollary 9.9. *$K[M]$ as an $M \times M$-module has a good filtration with composition factors of the form $H^0(\lambda) \otimes H^0(\lambda^*)$.*

Proof. This follows from c) using (2.2a) of [24].

We now explain the key features of the category $Rep(M)$ of rational M-modules.

a) $X(\overline{T})_+ = X(T)_+ \cap X(\overline{T})$ the **poset** of dominant weights of M.
b) $\{V_\lambda \mid \lambda \in X(\overline{T})_+\}$ the **simple** objects of $Rep(M)$.
c) $\{H^0(\lambda) \mid \lambda \in X(\overline{T})_+\}$ the **standard** objects of $Rep(M)$.

The following theorem is recorded in [26].

Theorem 9.10. *a) The socle of* $(H^0(\lambda))$ *is* V_λ *and the composition series of* $H^0(\lambda)/V_\lambda$ *has only factors of the form* V_μ *with* $\mu < \lambda$.
b) Each V_λ *has an injective hull* $V_\lambda \subseteq Q(\lambda)$ *so that* $Q(\lambda)$ *has a good filtration* $(0) = Q(\lambda)_0 \subseteq Q(\lambda)_1 \subseteq \ldots$ *with* $Q(\lambda)_1 = H^0(\lambda)$ *and* $Q(\lambda)_{i+1}/Q(\lambda)_i = H^0(\mu_i)$ *with* $\mu_i > \lambda$ *for* $i > 0$.

Proof. a) is well known, and b) follows from Proposition 9.8.

We have thus identified a key result about reductive normal monoids. $Rep(M)$ is a *highest weight category* in the sense of Cline, Parshall and Scott [16]. We cannot pursue all the important consequences of this result, but we shall give one striking illustration. Let M be reductive and normal, and suppose M has a zero element. Then we can write uniquely

$$K[M] = \bigoplus_{\chi \in Y} K[M]_\chi$$

where $Y = X(\overline{ZG^0})$, and each $K[M]_\chi$ is the subcoalgebra of $K[M]$ defined by

$$K[M]_\chi = \left\{ f \in K[M] \middle| \begin{matrix} f(gx) = \chi(g)f(x) \text{ for} \\ g \in \overline{ZG^\circ}, \text{ and } x \in M \end{matrix} \right\}.$$

It follows that each $K[M]_\chi$ is finite dimensional. Thus, for each $\chi \in Y$,

$$S(M)_\chi = Hom_K(K[M]_\chi, K)$$

is a finite dimensional K-algebra. It follows from Theorem 9.10 that

$$S(M)_\chi \text{ is quasihereditary.}$$

One could also prove this using Donkin's work since, in Donkin's notation, $S(M)_\chi$ is the generalized Schur algebra $S(\pi_\chi)$ where

$$\pi_\chi = i^{*-1}(\chi) \cap X(\overline{T})_+$$

and $i : \overline{Z(G)^\circ} \longrightarrow \overline{T}$ is the inclusion. Notice that $\pi_\chi \subseteq X(\overline{T})_+$ is a saturated subset of $X(\overline{T})$ with the given ordering.

9.3 The Blocks of $M_n(K)$ when $char(K) = p > 0$

A *block* can be thought of as an equivalence class of irreducible representations. The equivalence relation in this setup is generated by declaring irreducible representations (ρ, U) and (φ, W) to be in the same block if there exists an indecomposable representation (ψ, V) such that (ρ, U) and (φ, W) occur as factors in a composition series of (ψ, V). In this section we describe Donkin's calculation [25] of the blocks of $M_n(K)$. We end this section with a related, general conjecture about the blocks of irreducible, reductive monoids in characteristic $p > 0$.

Let $M = M_n(K)$, and let $T \subseteq G\ell_n(K)$ be the maximal torus of diagonal matrices. Then

$$X(T) = \mathbb{Z}^n = \{(\lambda_1, \ldots, \lambda_n) \mid \lambda_i \in \mathbb{Z}\},$$
$$X(\overline{T}) = \{(\lambda_1, \ldots, \lambda_n) \mid \lambda_i \geq 0\}$$

and

$$X(T)_+ = \{(\lambda_1, \ldots, \lambda_n) \mid \lambda_1 \geq \lambda_2 \geq \cdots \geq \lambda_n\} .$$

Thus

$$X(\overline{T})_+ = \{(\lambda_1, \ldots, \lambda_n) \mid \lambda_1 \geq \lambda_2 \geq \cdots \geq \lambda_n \geq 0\},$$

and from Theorem 9.6 we can identify the set of irreducible representations of M with $X(\overline{T})_+$ via the usual identification using highest weights $(V, \rho) = (V_\lambda, \rho_\lambda)$. So let $\lambda = (\lambda_1, \ldots, \lambda_n) \in X(\overline{T})_+$. Define

$$\sum_{i=1}^{n} \lambda_i = r, \text{ the } \textit{degree} \text{ of } \lambda,$$

and

$$d(\lambda) = max \left\{ d \geq 0 \,\middle|\, \begin{array}{l} \lambda_i - \lambda_{i+1} \equiv -1 \text{ mod}(p^d) \text{ for} \\ \text{all } i = 1, 2, \ldots n-1 \end{array} \right\} .$$

Assume that $char(K) = p > 0$.

Theorem 9.11 (Donkin's Theorem). *Let $(\rho_\lambda, V_\lambda)$ and (ρ_μ, V_μ) be irreducible representations of $M_n(K)$, where $\lambda = (\lambda_1, \ldots, \lambda_n)$ and $\mu = (\mu_1, \ldots, \mu_n)$. Then the following are equivalent.*

a) *λ and μ are in the same block.*
b) *i) λ and μ have the same degree.*
  ii) $d(\lambda) = d(\mu)$.
  iii) There exists $\pi \in S_n$ such that $\lambda_i - i \equiv \mu_{\pi(i)} - \pi(i) mod(p^{d+1})$ for all $i = 1, \ldots, n$.

The basic idea of the proof parallels that used earlier by Donkin in [23] for his classification of blocks of a semisimple algebraic group. Indeed, the first step is to reduce the problem to finding the blocks of the **Schur algebra** $S(n,r)$. The blocks of $S(n,r)$ are then just what one might suspect from knowing the blocks of $GL_n(K)$. But the latter may involve arbitrary r, and it is a delicate matter to compare the two situations.

There is a more conceptual way to state Donkin's Theorem that leads to an interesting conjecture. Let λ and μ be as above. By the results of [23], λ and μ are in the same block for $Gl_n(K)$ if and only if the three conditions of Theorem 9.11 b) above are satisfied. Furthermore, the blocks of any reductive group can be determined in the spirit of the above result for $Gl_n(K)$ using Theorem 5.8 of [23].

Let M be a reductive, normal monoid and let $\lambda, \mu \in X(\overline{T})_+$ represent irreducible representations of M (as in Theorem 9.6). Let

$$Bl_G(\lambda) = \{\nu \in X(T)_+ \mid \lambda \text{ and } \nu \text{ are in the same } G\text{-block}\}$$

and let

$$Bl_M(\lambda) = \{\nu \in X(\overline{T})_+ \mid \lambda \text{ and } \nu \text{ are in the same } M\text{-block}\}.$$

Conjecture 9.12. $Bl_M(\lambda) = Bl_G(\lambda) \bigcap X(\overline{T})_+$. *In particular,* λ *and* μ *are in the same* M*-block if and only if they are in the same* G*-block.*

9.4 The Blocks of Solvable Algebraic Monoids

It turns out that there is a striking description of the blocks of certain solvable algebraic monoids. While this is a special case, our basic idea here is expected to yield some decisive results for a large class of (nonsolvable) algebraic monoids in characteristic zero.

Let M be an irreducible, algebraic monoid with unit group G and zero element $0 \in M$.

Definition 9.13. *a)* M *is* solvable *if* G *is a solvable algebraic group (see Definition 2.33).*

b) M *is* polarizable *if* $0 \in \overline{Z(G)}$.

c) M *is* polarized *if we are given* $\theta : K^* \longrightarrow Z(G)$ *such that* θ *extends to* $\overline{\theta} : K \longrightarrow M$ *with* $\overline{\theta}(0) = 0$. *Then* θ *is called the* polarization, *and* (M, θ) *is a* polarized monoid.

It is easy to check that M is polarizable if and only if there exists a polarization $\theta : K^* \longrightarrow Z(G)$.

Let (M,θ) be polarized and solvable. Then $K^* \times M \longrightarrow M$, $(\alpha, x) \longmapsto \theta(\alpha)x$, induces a direct sum decomposition

$$\mathcal{O}(M) = \bigoplus_{n=0}^{\infty} \mathcal{O}_n(M)$$

where

$$\mathcal{O}_n(M) = \left\{ f \in \mathcal{O}(M) \,\middle|\, \begin{array}{l} f(\theta(\alpha)x) = \alpha^n f(x) \text{ for all} \\ \alpha \in K^*,\ x \in M \end{array} \right\} .$$

It follows easily that

$$\Delta(\mathcal{O}_n(M)) \subseteq \mathcal{O}_n(M) \otimes \mathcal{O}_n(M)$$

where Δ is the coalgebra structure on $\mathcal{O}(M)$.

Let (M, θ) and (N, ϕ) be polarized monoids. A θ-*morphism* between (M, θ) and (N, ϕ) is a morphism $\varphi : M \longrightarrow N$ of algebraic monoids such that $\varphi(0) = 0$ and $\varphi($ Image $(\theta)) \subseteq$ Image (ϕ). The θ-*degree* of φ is the degree of $\varphi|image(\theta)$. Then $\varphi(\theta(\alpha)) = \phi(\alpha^n)$ if φ is of θ-degree n.

Let A be a finite-dimensional associative K-algebra. The *blocks* of A are the obvious summands in the decomposition

$$A = \bigoplus_{e \in Z} eAe$$

where Z is the set of primitive, central idempotents of A. We denote the blocks of A by $B\ell(A)$ and identify them with Z.

If (M, θ) is a polarized monoid we define the **blocks** of M to be

$$B\ell(M) = \bigsqcup_{n \geq 0} B\ell(S_n(M))$$

where $S_n(M) := Hom_K(\mathcal{O}_n(M), K)$ has the algebra structure induced from the canonical coalgebra structure of $\mathcal{O}_n(M)$.

It is easy to check that this definition agrees with the one given by Green in 1.6b) of [33]. Indeed, he proves that any coalgebra (R, Δ) has a unique expression $R = \bigoplus_{\rho \in \mathcal{B}} R_\rho$ such that

i) $\Delta(R_\rho) \subseteq R_\rho \otimes R_\rho$ for all $\rho \in \mathcal{B}$,
ii) for any other direct sum decomposition $R = \bigoplus_{\gamma \in \Lambda} A_\gamma$ with $\Delta(A_\gamma) \subseteq A_\gamma \otimes A_\gamma$, each A_γ is a sum of some R_ρ's.

Proposition 9.14. *Let (M, θ) and (N, ϕ) be polarized monoids and let $\varphi : M \longrightarrow N$ be a dominant θ-morphism. Then φ induces a map of sets $B\ell(\varphi) : B\ell(N) \longrightarrow B\ell(M)$. This is a contravariant functor.*

Proof. If φ has θ-degree k then the induced morphism $\varphi^* : \mathcal{O}_n(N) \longrightarrow \mathcal{O}_{kn}(M)$ dualizes to obtain a surjective morphism of K-algebras

$$\varphi_n : S_{kn}(M) \longrightarrow S_n(N) .$$

For each primitive, central idempotent e of $S_n(N)$ there is a unique primitive, central idempotent $f \in S_{kn}(M)$ such that $\varphi_n(f)e = e$. Hence define $B\ell(\varphi)(e) = f$.

We now describe $B\ell(M)$ for a polarized monoid (M,θ) with solvable unit group. As it turns out, there is a straightforward description of $B\ell(M)$ in terms of weight spaces. As above we have $S_n(M) = Hom_K(\mathcal{O}_n(M), K)$. Define

$$\rho_n : M \longrightarrow S_n(M)$$

by $\rho_n(x)(f) = f(x)$. Then ρ_n is the universal θ-morphism of θ-degree n to a K-algebra. It follows easily that $\rho_n(M) \subseteq S_n(M)$ spans, and $S_n(M)$ is a solvable K-algebra. If $T \subseteq G$ is a maximal torus one checks that $D_n = span(\rho_n(T))$ is a maximal toral subalgebra of $S_n(M)$.

Proposition 9.15. *Let $T \subseteq G$ be a maximal torus and define $\mu_n : T \times T \times S_n(M) \longrightarrow S_n(M)$ by $\mu_n(s,t,x) = \rho_n(s)x\rho_n(t)$. Then there is a bijective correspondence between the nonzero weight spaces*

$${}^{\alpha}S_n^{\beta} = \{x \in S_n(M) \mid \rho_n(s)x\rho_n(t) = \alpha(s)\beta(t)x \text{ for all } s,t \in T\}$$

and the pairs of primitive idempotents $(e,f) \in E(D_n)\times E(D_n)$ with $eS_n(M)f \neq (0)$. Also, (e,f) corresponds to the unique ${}^{\alpha}S_n^{\beta}$ with $eS_n(M)f = {}^{\alpha}S_n^{\beta}$.

Proof. We leave the details to the reader. The proof hinges on identifying the set of primitive idempotents of D_n with characters of T. See Proposition 2.3 of [113].

Thus we define

$$S = \{(\alpha,\beta) \in X(T) \times X(T) \mid {}^{\alpha}S_n^{\beta} \neq 0 \text{ for some } n \geq 0\}$$

$$\Delta(\overline{T}) = \{(\alpha,\beta) \in X(\overline{T}) \times X(\overline{T}) \mid \alpha = \beta\} .$$

One checks that

$$\Delta(\overline{T}) \subseteq S \subseteq X(\overline{T}) \times X(\overline{T}) .$$

We use S to define an equivalence relation on $X(\overline{T})$. For $\alpha, \beta \in X(\overline{T})$ we define

$$\alpha \longrightarrow \beta \text{ if } (\alpha,\beta) \in S, \text{ and}$$
$$\alpha \longleftarrow \beta \text{ if } (\beta,\alpha) \in S .$$

Lemma 9.16. *a) The equivalence relation on $X(\overline{T})$ generated by $\longrightarrow$ is the same as the equivalence relation generated by $\longleftarrow$. It can be described as follows:*

$$\alpha \sim \beta \textit{ if there exist } \gamma_1, \ldots, \gamma_{2m-1} \in X(\overline{T})$$

such that

$$\alpha = \gamma_1 \longrightarrow \gamma_2 \longleftarrow \ldots \longleftarrow \gamma_{2m-1} = \beta.$$

b) Suppose that $\alpha \sim \beta$ and $\lambda \sim \delta$. Then $\alpha\lambda \sim \beta\delta$.

Proof. a) is a straightforward calculation. For b) one uses the fact that S is a semigroup, together with the fact that the "m" for $\alpha \sim \beta$ can be chosen equal to the "m" for $\lambda \sim \delta$.

Theorem 9.17. $(X(\overline{T})/\sim) \cong B\ell(M)$ *via* $\alpha \longmapsto [\alpha]$. *Then* $X(\overline{T}) \longrightarrow B\ell(M)$ *is a surjective morphism of monoids.*

Proof. $B\ell(M) = \bigsqcup_{n\geq 0} B\ell(S_n(M))$. Hence let $D_n = Span(\rho_n(T))$ be the maximal toral subalgebra as discussed above. Using standard facts about associative algebras we see that

$$B\ell(S_n(M)) = E_1(D_n)/\sim_\circ$$

where $\sim_\circ$ is the equivalence relation on $E_1(D_n)$ generated by declaring

$$e \sim_\circ f \quad \text{if} \quad eS_n(M)f \neq 0 \quad \text{or} \quad fS_n(M)e \neq 0\,.$$

But $E_1(D_n)$ is identified, via Proposition 9.15, with $S_n = \{(\alpha,\beta) \in S \mid {}^{\alpha}S_n(M)^{\beta} \neq 0\}$. Thus the two equivalence relations correspond. We conclude that $X(\overline{T})/\sim \cong \left(\bigsqcup_{n\geq 0} E_1(D_n)\right)/\sim_\circ$.

Now $X(\overline{T}) \longrightarrow X(\overline{T})/\sim$ determines a subscheme $Y \subseteq \overline{T}$ via

$$Y = Spec(K[X(\overline{T})/\sim]) \qquad \text{(monoid algebra)}$$

and one obtains, from 3.2 of [113], that $Y \subseteq Z(M)$, the center of M. Furthermore, the surjection, $K[M] \longrightarrow K[Y]$, identifies $S_n(Y)$ with the maximal toral subalgebra of $Z(S_n(M))$. In particular,

$$Y = \bigcap_{g\in G} g\overline{T}g^{-1}\,.$$

From these comments, and a little more calculation (3.5 of [113]), we obtain the following theorem.

Theorem 9.18. *There is a canonical bijection*

$$Bl_n(M,\theta) \cong X_n(Y)$$

where $X_n(Y) = \{\chi : Y \longrightarrow K \mid \chi$ *has* θ*-degree* $n\}$.

We conclude the chapter with three examples and a conjecture.

Example 9.19. We define polarizable, solvable monoids M and M' as follows:

$$M = \{(u,(r,s)) \mid u,r,s \in K\}$$

with multiplication

$$(u,(r,s))(v,(k,\ell)) = (k\ell u + r^2 v,\ (rk, s\ell)),$$

and

$$M' = \{(u,(r,s) \mid u,r,s \in K\}$$

with multiplication

$$(u,(r,s))(v,(k,\ell)) = (\ell u + rv,\ (rk, s\ell)).$$

Define $\varphi : M' \longrightarrow M$ by $\varphi(u,(r,s)) = (ru,(r,s))$. One checks that φ is a birational θ-morphism of degree one. Furthermore, φ induces an isomorphism

$$\varphi : \overline{T}' \xrightarrow[\cong]{} \overline{T}\,.$$

We now compute the center of each monoid. Clearly,

$$Z(M) = \{(0,(r,r)) | r \in K\}$$

since M is the algebra of 2×2 upper triangular matrices. As for M', one needs a little more calculation, and we obtain

$$Z(M') = \{(0,(r,r)) \mid r \in K\} \cup \{(0,(0,s)) \mid s \in K\}\,.$$

In particular, the inclusion $Z(M) \subset Z(M')$ is proper, so that M and M' have different block structure even though $\varphi : M' \longrightarrow M$ is a birational equivalence with $\overline{T}' \xrightarrow[\cong]{} \overline{T}$.

Example 9.20. Define a polarizable solvable monoid N as follows:

$$N = \{(u,(r,s)) \mid u,r,s \in K\}$$

with multiplication

$$(u,(r,s))(v,(k,\ell)) = (k^2\ell u + r^3 v,\ (rk, s\ell)).$$

One checks that $\overline{T} = \{(0,(\alpha,\beta)) | \alpha,\beta \in K\}$ is the closure in N of the maximal torus $T = \{(0,(\alpha,\beta)) \mid \alpha\beta \neq 0\}$. Hence assume $(0,(r,s)) \in \overline{T}$ is central. Then we must have

$$(0,(r,s))(v,(k,\ell)) = (v,(k,\ell))(0,(r,s)) \qquad \text{for all } v,k,\ell\,.$$

Thus $r^3 v = sr^2 v$ for all v and so $r^3 = sr^2$. By the comment preceding Theorem 9.18

$$\begin{aligned} K[Y] &= K[U,R,S]\big/(U, R^3 - SR^2) \\ &\cong K[R,S]\big/(R^3 - SR^2)\,. \end{aligned}$$

Let $x = \overline{R}$, $y = \overline{S} \in K[y]$. Then $f = x(x-y) \neq 0$, and yet

$$f^2 = x^2(x-y)^2 = (x^3 - x^2y)(x-y) = 0 \, .$$

Thus $K[Y]$ is not reduced. We can also calculate the number of blocks of N of each θ-degree, using this presentation of $K[Y]$. In fact,

$$\begin{aligned} |B\ell_0(N)| &= 1 \\ |B\ell_1(N)| &= 2 \\ |B\ell_n(N)| &= 3 \quad \text{if } n \geq 2 \, . \end{aligned}$$

It appears that there may be an important structural relationship between the blocks of solvable monoids, and the blocks of arbitrary irreducible monoids (at least in characteristic zero).

Example 9.21. We start with $M = G$, a semisimple, simply connected algebraic group of rank r in characteristic zero. Here we find that

$$Bl(G) = IR(G),$$

and this is a monoid under the "Cartan product". Let $T \subset B \subset G$ be a maximal torus of the Borel subgroup B of G. Let $\{\rho_1, \dots, \rho_r\}$ be the set of fundamental, dominant representations of G, so that $\rho_i : G \longrightarrow Gl(V_i)$. Let $L_i \subset V_i$ be the unique, one-dimensional subspace stabilized by B. Let $D_n(K)$ be the monoid of diagonal $n \times n$ matrices. Define

$$\varphi : T \longrightarrow D_n(K)$$

by $\varphi(t) = (\rho_1(t)|L_1, \dots, \rho_r(t)|L_r)$. Then φ is the restriction of a $U \times U^-$-equivariant morphism $\psi : G \longrightarrow D_r(K)$, which is defined on UTU^- by $\psi(utv) = \varphi(t)$. Somehow, the $U \times U^-$-morphism ψ might be thought of as a kind of "basic monoid" associated with G. In particular,

a) $D_r(K)$ is solvable,
b) $Bl(G) \cong Bl(D_r(K))$,
c) G and $D_n(K)$ are (somehow) Morita equivalent via ψ.

Notice that even though G is a group, this "basic" object $D_n(K)$ is a monoid.

The above example leads us to an interesting conjecture about the blocks of algebraic monoids.

Conjecture 9.22. Let M be an irreducible, algebraic monoid. There exists an irreducible, algebraic monoid $B(M)$, and a certain dominant morphism $\psi : M \longrightarrow B(M)$ of algebraic varieties, such that

a) $B(M)$ is solvable,
b) M and $B(M)$ are (somehow) Morita equivalent via ψ
c) in particular, $Bl(M) \cong Bl(B(M))$.

The reader might wonder what it means for algebraic monoids M and N to be *Morita equivalent.* Unfortunately, this has yet to be formulated precisely. Obviously, it will involve a bijection between $Bl(M)$ and $Bl(N)$, as well as a diagram of Morita equivalences between the corresponding (block) algebras one obtains from the "coordinate coalgebras" of M and N. In any case, there should be enough clues in the above example to find the "correct" definition, at least in the case of polarized monoids.

10

Monoids of Lie Type

Algebraic monoids are rich in structure, mainly because they are algebraic varieties with much symmetry. In this chapter we focus on those properties that allow us to identify what makes the theory "tick" from an abstract semigroup point of view. The reader should think of this development as a natural extension of Tits' viewpoint (in [140]) to the case of monoids. One might even hope that the theory of spherical embeddings is the undisputed clue that will someday lead us to the ultimate formulation of abstract, combinatorial diagram geometry.

10.1 Finite Groups of Lie Type

The classification of reductive, normal monoids is independent of the (algebraically closed) ground field (see Theorem 5.4). Thus it can be applied uniformly to the algebraic closure of finite fields. This allows us to define a class of finite monoids in the spirit of Steinberg's theory of Chevalley groups and their twisted analogues [133]. Such finite monoids are useful in counting problems related to the Weil zeta function.

Let G be a simple, algebraic group defined over a field of characteristic $p > 0$. Chevalley classified all endomorphisms $\sigma : G \longrightarrow G$ with the property that

$$G_\sigma := \{x \in G \mid \sigma(x) = x\}$$

is a finite group. There are essentially two types.

10.1.1 σ Preserves Root Length

Then σ induces an automorphism ρ of the root system of G so that $\sigma(t) = Fr_q(\rho(t))$, where $Fr_q(t) = t^q$ for all $t \in T$, an appropriately chosen maximal torus of G. The root systems with automorphisms are A_n, D_n and E_6. It turns out that $Aut(A_n) = \mathbb{Z}/2\mathbb{Z}$, $Aut(D_n) = \mathbb{Z}/2\mathbb{Z}$ $(n > 4)$, $Aut(D_4) = S_3$ and $Aut(E_6) = \mathbb{Z}/2\mathbb{Z}$.

10.1.2 σ Exchanges Root Length

Then G is of the type C_2, F_4 or G_2, and $char(K) = 2$ or 3. See Chapter II of [133] for more details.

Chevalley ($\rho = id$) and Steinberg ($\rho \neq id$) studied the groups of type 10.1.1, and Suzuki (C_2) and Ree (F_4, G_2) studied the groups of type 10.1.2. Later Tits [140] found the method (BN pairs) that accommodates the structure theory of all sixteen families.

Altogether, there are sixteen different families. In addition to those families where ρ is the identity, there are seven other families:

ρ as in 10.1.1: $A_n^2, D_n^2, D_4^3, E_6^2$;
ρ as in 10.1.2: C_2^2, F_4^2, G_2^2.

These sixteen families, together with the prime cyclic groups and the alternating groups for $n \geq 5$, account for all but twenty six of the finite simple groups.

We are interested in identifying the endomorphisms that yield finite fixed point monoids with good behaviour. The following result does the job.

Let G be a connected algebraic group, and let σ be an algebraic group endomorphism of G. Denote by $1 - \sigma : G \to G$, the morphism of varieties $1 - \sigma(g) = g\sigma(g)^{-1}$. As above, let

$$G_\sigma := \{x \in G \mid \sigma(x) = x\}.$$

Theorem 10.1. (Lang's Theorem) *Let G and σ be as above, and assume that σ is surjective. If G_σ is finite then $(1 - \sigma)(G) = G$. If G is semisimple, then G_σ is finite if and only if the differential $d\sigma$ of σ is nilpotent.*

See Theorems 10.1 and 10.5 of [133].

10.2 Endomorphisms of Linear Algebraic Monoids

Theorem 10.2. *Let $\sigma : M \longrightarrow M$ be an endomorphism of the reductive monoid M. The following are equivalent:*

a) σ is a finite morphism and $G_\sigma = \{x \in G \mid \sigma(x) = x\}$ is a finite group;
b) The morphism $1 - \sigma : G \longrightarrow G$, $(1 - \sigma)(x) = x\sigma(x)^{-1}$, is surjective, and there exists a maximal torus $T \subseteq G$ so that $\sigma(\overline{T}) = \overline{T}$.

The keys to Theorem 10.2 are Lang's Theorem (10.1 above), to get $1 - \sigma$ surjective when G_σ is finite; and Theorem 4.2, to get σ finite when $\sigma(\overline{T}) = \overline{T}$.

The importance of Theorem 10.2 will become apparent in Theorem 10.4.

We define, for σ and M as in 10.2,

$$M_\sigma = \{x \in M \mid \sigma(x) = x\} .$$

To obtain the structure of M_σ we need to know a little more about the homogeneity properties of this σ-process. The following result is due to Springer and Steinberg (Section E of [131]).

Proposition 10.3. *Let $G \times X \to X$ be a homogeneous space for the connected group G. Assume that $\sigma : G \to G$ is such that G_σ is finite. Suppose that $\tau : X \to X$ is a morphism compatible with $\sigma : G \to G$ (so that $\tau(gx) = \sigma(g)\tau(x)$). Then $X_\tau = \{x \in X \mid \tau(x) = x\} \neq \phi$. If G_x is connected for $x \in X$, then $G_\sigma \times X_\tau \to X_\tau$ is transitive.*

The proof of Proposition 10.3 is a straightforward application of Lang's Theorem.

Applying Proposition 10.3 to the appropriate orbits of G or $G \times G$ on the reductive monoid M yields the following result. See Theorem 4.3 of [96] for the details.

Theorem 10.4. *Let M be reductive with $\sigma : M \longrightarrow M$ as in Theorem 10.2.*

a) M_σ is finite.
b) M_σ is unit regular; $M_\sigma = E(M_\sigma)G_\sigma$.
c) If $T \subseteq B$ satisfies $\sigma(T) = T$ and $\sigma(B) = B$, then $M_\sigma = \bigcup_{e \in \Lambda_\sigma} G_\sigma e G_\sigma$ where $\Lambda_\sigma = \{e \in E(\overline{T}) \mid Be = eBe \text{ and } \sigma(e) = e\}$. Such T and B exist by the results of [133].
d) If $e \in E_\sigma$ then $P_\sigma(e)$ and $P_\sigma^-(e)$ are opposite parabolic subgroups of G_σ.
e) If $e, f \in E_\sigma$ and $eM_\sigma = fM_\sigma$ or $M_\sigma e = M_\sigma f$, then there exists $g \in G_\sigma$ such that $geg^{-1} = f$.
f) If $e \in \Lambda_\sigma$ and $U \lhd P_\sigma(e)$ is the unipotent radical, then $Ue = \{e\}$. Similarly for $U^- \lhd P_\sigma^-(e)$, $eU^- = \{e\}$.

Putcha calls such monoids, satisfying a) - f), *monoids of Lie type.* He uses this as the starting point for his theory of "monoids on groups with BN pairs". See § 10.4 for the details of this surprising development, including a classification theory based on type maps.

10.3 A Detailed Example

The purpose of this example is to illustrate the two significant features while comparing M and M_σ.

a) The BN pair structure of G and G_σ can be different, i.e. G and G_σ can be associated with different Dynkin diagrams.
b) It can happen that $\Lambda_\sigma \subsetneq \Lambda$.

Actually a) and b) are equivalent for semisimple monoids.

Let $K = \overline{\mathbb{F}_q}$ and let $G = S\ell_4(K) \times K^*$. Define $\sigma : G \longrightarrow G$ by

$$\sigma(x, \alpha) = (Fr_q(w\, T(x^{-1})w), \alpha^q)$$

where $Fr_q((x_{ij})) = (x_{ij}^q)$ and $w = \begin{pmatrix} 0\,0\,0\,1 \\ 0\,0\,1\,0 \\ 0\,1\,0\,0 \\ 1\,0\,0\,0 \end{pmatrix}$, and T is the transpose operator.

The group G_σ here is a central extension of a unitary group over $\mathbb{F}_q$. See [135] for those details. The diagonal maximal torus T of G has character group $X(T)$ with presentation

$$X(T) = \langle x_1, x_2, x_3, x_4, \delta \mid x_1 + x_2 + x_3 + x_4 = 0 \rangle$$

where $\chi_i\,((\alpha_1, \alpha_2, \alpha_3, \alpha_4, \alpha)) = \alpha_i$ and $\delta(\alpha_1, \alpha_2, \alpha_3, \alpha_3, \alpha_4, \alpha) = \alpha$. One checks that $\sigma^* : X(T) \longrightarrow X(T)$ is given by

$$\begin{aligned} \sigma^*(\chi_1) &= -q\chi_4 \\ \sigma^*(\chi_2) &= -q\chi_3 \\ \sigma^*(\chi_3) &= -q\chi_2 \\ \sigma^*(\chi_4) &= -q\chi_1 \\ \sigma^*(\delta) &= q\delta \,. \end{aligned}$$

Hence

$$\begin{aligned} \sigma^*(\chi_1 - \chi_2) &= q(\chi_3 - \chi_4) \\ \sigma^*(\chi_2 - \chi_3) &= q(\chi_2 - \chi_3) \\ \sigma^*(\chi_3 - \chi_4) &= q(\chi_1 - \chi_2) \,. \end{aligned}$$

In the notation of 7.1, ρ^* is given on the base $\Delta = \{\chi_1 - \chi_2,\ \chi_2 - \chi_3,\ \chi_3 - \chi_4\}$ via

$$\begin{aligned} \rho(\chi_1 - \chi_2) &= \chi_3 - \chi_4 \\ \rho(\chi_2 - \chi_3) &= \chi_2 - \chi_3 \\ \rho(\chi_3 - \chi_4) &= \chi_1 - \chi_2 \,, \end{aligned}$$

and $\sigma = Fr_q \circ \rho$. Hence σ is of type A_3^2.

We now construct an interesting reductive monoid M with unit group G so that $\sigma : G \longrightarrow G$ extends to an endomorphism $\sigma : M \longrightarrow M$ that satisfies the conditions of Theorem 10.2. To construct M we use Theorem 5.4. Then it suffices to find a finitely generated submonoid $C \subseteq X(T)$ such that

i) C generates $X(T)$ as a group
ii) C is invariant under the Weyl group

iii) $C = \{x \in C \mid n\chi \in C \text{ for all } n > 0\}$
iv) $qC \subseteq \sigma^*(C) \subseteq C$.

To do this we consider

$$C_1 = \langle \chi_i - \chi_j + \delta \mid i \neq j \rangle$$

and we let

$$C = \{\chi \in X(T) \mid n\chi \in C_1\} .$$

Then by Theorem 5.4 and Theorem 5.2 there is a unique, normal monoid M such that

i) $G = Sl_4(R) \times K^*$
ii) if $T \subseteq G$ is a maximal torus then $X(\overline{T}) = C$
iii) the morphism $\sigma : G \longrightarrow G$ extends to a finite, surjective morphism $\sigma : M \longrightarrow M$.

A more direct construction of M can be obtained as follows. Let $\mathfrak{g}$ be the Lie algebra of $S\ell_4(K)$, and let $\rho : G \longrightarrow Aut(\mathfrak{g})$ be defined by $\rho(g, \alpha) = \alpha Ad(g)$, where Ad is the adjoint representation of $S\ell_4(K)$. Then there is a finite dominant morphism from M to the Zariski closure of the image of ρ.

The following properties of M are easily obtained.

a) M has a unique minimal, nonzero $G \times G$-orbit corresponding to the Weyl group orbit of $\chi_1 - \chi_4 + \delta$ (see § 7.2)
b) $\rho(\chi_1 - \chi_4 + \delta) = \chi_1 - \chi_4 + \delta$
c) M is the $\mathfrak{J}$-irreducible monoid of type $J_0 = \{\beta\}$ where the Dynkin diagram is

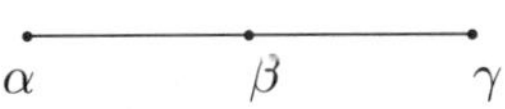

Thus, by Theorem 7.12 a), the cross-section lattice is as follows:

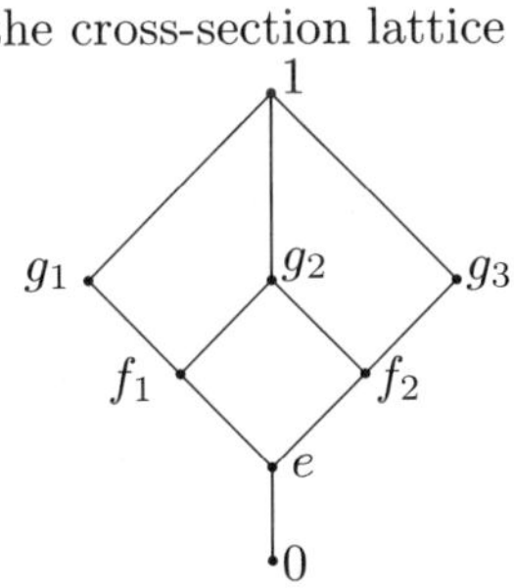

where $\Lambda \equiv \{\phi, g_1, g_2, g_3, f_1, f_2, e, 0\}$. The type map is given as follows:

$$\begin{aligned}
\lambda(0) &= \{\alpha, \beta, \gamma\} \\
\lambda(e) &= \{\beta\} \\
\lambda(f_1) &= \{\alpha\} \\
\lambda(f_2) &= \{\gamma\} \\
\lambda(g_1) &= \{\alpha, \beta\} \\
\lambda(g_2) &= \{\alpha, \gamma\} \\
\lambda(g_3) &= \{\beta, \gamma\} \\
\lambda(1) &= \{\alpha, \beta, \gamma\} = S \,.
\end{aligned}$$

Also, from the above calculation of σ^* we obtain $\sigma(e) = e$, $\sigma(f_1) = f_2$, $\sigma(f_2) = f_1$, $\sigma(g_1) = g_3$, $\sigma(g_3) = g_1$ and $\sigma(g_2) = g_2$. By 10.4, $M_\sigma = \bigsqcup_{f \in \Lambda_\sigma} G_\sigma f G_\sigma$, where

$$\Lambda_\sigma = \{0, e, g_2, 1\} \,.$$

The group G_σ is a finite group with BN-pair associated with the diagram

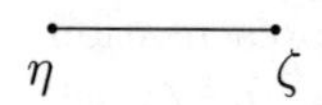

obtained by "folding" the diagram

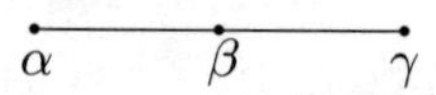

at β, thereby identifying α and γ. The type map of M_σ is given by

$$\begin{aligned}
\lambda(e) &= \{\zeta\} \longleftrightarrow \{\beta\} \\
\lambda(g_2) &= \{\zeta\} \longleftrightarrow \{\alpha, \gamma\} \,.
\end{aligned}$$

10.4 Abstract Monoids of Lie Type

In this section, we describe Putcha's theory of abstract monoids of Lie type. This was developed by Putcha in [85, 86] as the method of understanding many of the purely semigroup theoretic aspects of the theory of algebraic monoids. The reader might consider this development as the natural monoid analogue of Tits' theory of groups with BN-pair.

In this section, we refer the reader to [85, 86, 97] for many of the proofs.

Definition 10.5. *Let M be a regular monoid with unit group G and idempotent set $E(M)$. Assume that G is a group with BN-pair (B, N, S, W). We say that M is a* monoid of Lie type *on G if*

a) M *is generated by* G *and* $E(M)$
b) *If* $e \in E(M)$, *then* $P(e) = \{g \in G \mid ge = ege\}$ *and* $P^-(e) = \{g \in G \mid eg = ege\}$ *are opposite parabolic subgroups of* G, *and* $eR_u(P^-(e)) = \{e\} = R_u(P(e))e$
c) *If* $e, f \in E(M)$, *and* $e\mathfrak{R}f$ *or* $e\mathfrak{L}f$, *then there exists* $g \in G$ *such that* $geg^{-1} = f$.

It follows easily that if $ef = fe = f$ then $R_u(P(e)) \subseteq P(f)$, and so $P(e, f) = P(e) \cap P(f)$ is a parabolic subgroup of G. Furthermore, $P(e, f)$ and $P^-(e, f)$ are opposite parabolic subgroups.

The following basic result is obtained in [85]. Let B be a Borel subgroup of G. We say that P and P^- are **standard** opposite parabolic subgroups if $B \subseteq P$.

Theorem 10.6. *Let* M *be a monoid of Lie type on* G. *Then*

a) $M = E(M)G = E(M)G$
b) *the partially ordered set* $\mathfrak{U}$ *of* $\mathfrak{J}$*-classes of* M *is a lattice*
c) *for each* $J \in \mathfrak{U}$ *there is a unique* $e_J \in E(J)$ *such that* $P(e)$ *and* $P^-(e)$ *are standard opposite Borel subgroups*
d) *if* $J_1, J_2 \subseteq \mathfrak{U}$ *then* $e_{J_1}e_{J_2} = e_{J_1 \wedge J_2}$, *and hence* $\Lambda = \{e_J \mid J \in \mathfrak{U}\} \cong \mathfrak{U}$ *as lattices.*

Just as in the case of reductive monoids, we refer to Λ as the **cross-section lattice** of M. Furthermore, we have the **type map**

$$\lambda : \Lambda \to 2^S$$

defined by setting $\lambda(e_J) = C_S(e_J) = \{s \in S \mid se_J = e_J s\}$.

Remark 10.7. There is a purely combinatorial characterization of the type maps of monoids of Lie type. We refer the reader to [85].

It is useful to know how one constructs a monoid of Lie type directly from a type map of the form mentioned in Remark 10.7 above. In the following example, we describe how this is done for a canonical monoid. The details of this construction are recorded in [66, 96].

Example 10.8. Let G be a group of Lie type with BN-pair structure (G, B, N, S). If $I \subseteq S$ let P_I be the standard parabolic subgroup of G of type I, and let P_I^- be the parabolic subgroup opposite to P relative to $T = N \cap B$. It is well known that

$$P_I^- P_I \cong U_I^- \times L_I \times U_I.$$

where $P_I \cap P_I^- = L_I$ is the Levi factor common to P_I and P_I^-. Thus we have a natural projection map

$$\theta_I : P_I^- P_I \to L_I.$$

Restricted to P_I or P_I^-, θ_I is a homomorphism.

For each $I \subseteq S$, define an idempotent e_I so that

$$\Lambda = \{e_I \mid I \subseteq S\} \cup \{0\}$$

and $e_I \wedge e_K = e_{I \cap K}$ if $I \cap K \neq \phi$, and zero otherwise. For $e_I \in \Lambda$, define

$$J_I = Ge_IG/\sim$$

where $xe_Iy \sim ue_Iv$ if $u^{-1}x \in P_I$, $vy^{-1} \in P_I^-$ and $\theta(u^{-1}x) = \theta(vy^{-1})$. Finally, define

$$M = \bigsqcup_{I \subseteq S} J_I \sqcup \{0\}.$$

We now define the multiplicative structure on M. Let $a = xe_Iy$ and $b = se_Kt$. Then

$$ab = \begin{cases} 0 & \text{, if } ys \notin P_I^- P_K \\ xle_{I\cap K}mt & \text{, if } ys \in U_I^- lmU_K,\ l \in L_I,\ m \in L_K\ . \end{cases}$$

It turns out that M is a monoid of Lie type with type map

$$\lambda : \Lambda \to 2^S$$

defined by $\lambda(e_I) = I$, $\lambda(0) = S$. Hence M is a *canonical monoid.*

The next problem is to quantify the extent to which a monoid of Lie type is determined by its unit group G, and its type map $\lambda : \Lambda \to 2^S$. There is a very satisfying answer ([85]), which we now describe.

Let M be a monoid of Lie type, and let $J \in \mathfrak{U}$. Define

$$K_J = \{g \in G \mid ge_J = e_Jg = e_J\}.$$

Then K_J is a normal subgroup of $L_J = P(e) \cap P^-(e)$. Let

$$\mathcal{K} = \{(J, K_J) \mid J \in \mathfrak{U}\}.$$

Let M_1 and M_2 be monoids of Lie type with unit group G, and the same type map λ. If $\mathcal{K}_1 = \mathcal{K}(M_1)$ and $\mathcal{K}_2 = \mathcal{K}(M_2)$, we say that

$$\mathcal{K}_1 \geq \mathcal{K}_2$$

if for all $J \in \mathfrak{U}$, $(K_1)_J \subseteq (K_2)_J$.

Theorem 10.9. *Let M be a monoid of Lie type with unit group G and type map $\lambda : \Lambda \to 2^S$.*

a) M is completely determinied by $(\lambda, \mathcal{K}(M))$.

b) *Let M_1 and M_2 be monoids of Lie type with the same unit group G and the same type map λ. There is a morphism of monoids $M_1 \to M_2$, extending the identity map on $\Lambda \cup G$, if and only if $\mathcal{K}_1 \geq \mathcal{K}_2$.*

c) *There is a largest monoid $M^+(\lambda)$ of type λ with $K_G = \{1\}$. $\mathcal{K}(M^+(\lambda)) = \{K_J^+ \mid J \in \mathcal{U}\}$, where K_J^+ is the subgroup of L_J generated by all U_H and U_H^-, where $H \in \mathcal{U}$ and $H \geq J$. In particular, there is a unique morphism $M^+(\lambda) \to M$, extending the identity on $\Lambda \cup G$, for any monoid of Lie type M with group G and type λ.*

d) *There is a smallest monoid $M = M^-(\lambda)$ of type λ with $K_G = \{1\}$. $\mathcal{K}(M^-(\lambda)) = \{K_J^- \mid J \in \mathcal{U}\}$ where*

$$K_J^- = \bigcap_{g \in L_J, J \geq H} g(L_H \cap L_J)g^{-1}.$$

In particular, there is a unique morphism $M \to M^-(\lambda)$, extending the identity on $\Lambda \cup G$, for any monoid of Lie type M with group G and type λ.

Parts a), b) and d) of Theorem 10.9 are proved in [85], and part c) is proved in [86]. The reader should think of the type map here as determining M to within a kind of "central extension of monoids". In the geometric case, it is entirely likely that the isomorphism classes of monoids of fixed type could be organized into families of commutative semigroups.

A more refined description of K_J^+ and K_J^- has since been given in Theorem 1.1 of [97]. In particular, both K_J^+ and K_J^- have been identified in terms of certain Levi subgroups of standard parabolics.

Let M be a finite monoid of Lie type. Let G be the group of units of M, and let Λ be a cross-section lattice of M. Recall from § 2.3.1, the notion of *congruence* on M. A congruence on M arising from a congruence on Λ is called a *discrete* congruence. At the other extreme, a congruence on M that arises from an idempotent separating homomorphism is called an *idempotent separating* congruence. It is shown in [3] that any congruence on M, which is the identity on G, factors as a discrete congruence followed by an idempotent separating congruence.

10.5 Modular Representations of Finite Reductive Monoids

In this section we consider reductive algebraic monoids M defined over a finite field $k = \mathbb{F}_q$. We readily obtain finite reductive monoids M_r, $r \geq 1$, as follows.

Let $M_r = M(\mathbb{F}_{q^r})$ be the finite monoid of $\mathbb{F}_{q^r}$-rational points of M. By standard facts about finite fields and Galois groups (Chapter II, Section 4 of [61]) there exists an $\mathbb{F}_q$-automorphism $\sigma : M \longrightarrow M$ of algebraic monoids such that

$$M_r = \{x \in M \mid \sigma^r(x) = x\}\ .$$

Thus, M_r is a *monoid of Lie type* in the sense of § 10.4. Such monoids enjoy many special properties.

In this section we describe a useful formula for the number $|IR(M_r)|$ of irreducible, modular representations of M_r. The basic problem here is to consider the fomulas of the form

$$|IR(M_r)| = (q^r - 1)\sum_{i=1}^{n} a_i q^{ri}$$

where $a_i \in \mathbb{Z}$ and is independent of r. Whenever this can be done, it is particularly interesting to interpret the a_i.

We now state the main result of [96]. This is the main reason we are able to obtain so much information about irreducible modular representations of finite monoids of Lie type. Let M be a finite monoid of Lie type with unit group G of characteristic p.

Theorem 10.10. *Suppose that $\rho : M \to End(V)$ is an irreducible representation of M over $\overline{\mathbb{F}}_p$. Then $\rho|G$ is irreducible.*

Proof. First we consider the special case $M = M(G)$, where $M(G)$ is the *canonical monoid* of Example 10.17 below. It is then possible to construct

$$\sum_{I \subseteq S} 2^{|S\setminus I|}\alpha_I$$

inequivalent, irreducible $M(G)$-modules, each of which restricts to an irreducible representation of G. Here, $\alpha_I = |Hom(L_I, \overline{\mathbb{F}}_p)|$, where L_I is a Levi factor of P_I. On the other hand, the number of irreducible representations of $M(G)$ is

$$\sum_{I \subseteq S}\ \sum_{K \subseteq I} \alpha_K .$$

Since the two numbers above are equal, the theorem is true for $M(G)$.

To get the result for any finite monoid M of Lie type we start with an irreducible representation $\rho : M \longrightarrow End(V)$ over $\overline{\mathbb{F}}_p$. It is then possible to construct an irreducible representation, as in Corollary 2.7 of [96], $\overline{\rho} : M(G) \longrightarrow End(V)$ such that $\rho|G = \overline{\rho}|G$. Thus, the theorem is proved for M.

The reader is reminded here that there is rarely such a direct and appealing relationship between the irreducible representations of a monoid and those of its unit group.

Let M be a finite monoid of Lie type with unit group G. Let S be the Coxeter-Dynkin diagram of G. Let $\mathfrak{U} = \mathfrak{U}(M)$ be the set of regular $\mathfrak{J}$-classes of M. It turns out that $\mathfrak{U}$ is the set of two-sided G-orbits of M. Define $GaG \geq GbG$ if $b \in MaM$. In this way, $\mathfrak{U}$ becomes a lattice. There is a cross-section of idempotents $\Lambda = \{e_J | J \in \mathfrak{U}\} \subseteq E(M)$, such that $J = Ge_JG$ and, for all $J_1, J_2 \in \mathfrak{U}, e_{J_1}e_{J_2} = e_{J_2}e_{J_1} = e_{J_1 \wedge J_2}$. Then Λ is called a *cross section lattice.* Furthermore, $E(J) = \{ge_Jg^{-1} | g \in G\}$.

Recall from Definition 4.6 b), the *type map*:

$$\lambda : \Lambda \to 2^S.$$

It is defined so that for all $J \in \mathfrak{U}$

$$P(e_J) = P_{\lambda(J)}.$$

$P(e) = C_G^r(e) =: \{g \in G | ge = ege\}$, and $P_I \subseteq G$ denotes the *parabolic subgroup of type* I as in Theorem 2.44.

Let M be a finite monoid of Lie type of characteristic p, and let $\rho : M \to End(V)$ be an irreducible representation of M defined over $\overline{\mathbb{F}}_p$. By the theory of Munn and Ponizovskii [15], ρ determines an *apex*, $Apex(\rho) \in \mathfrak{U}(M)$. By definition $Apex(\rho)$ is the unique smallest $\mathcal{J}$-class J of M such that $\rho(J) \neq 0$. But on the other hand, $\rho|G$ is irreducible by Theorem 10.10 above, and so by the theory of Richen [18] $\rho|G$ is determined by its *weight* $(I(\rho), \chi(\rho))$. In any case, $\rho : M \to End(V)$ determines the following data:

(i) $J = Apex(\rho) \in \mathfrak{U}(M)$
(ii) $I = I(\rho) \in 2^S$
(iii) $\chi = \chi(\rho) : P_I \to \overline{\mathbb{F}}_p^*$.

We consider the following two questions.

(a) Is ρ uniquely determined up to equivalence of representations by (J, I, χ)?
(b) What are the conditions on a triple (J, I, χ) with $\chi : P_I \to \overline{\mathbb{F}}_p^*$ and $J \in \mathfrak{U}(M)$, so that there exists an irreducible representation ρ of M with
  (i) $\rho|G$ of type (I, χ)
  (ii) $Apex(\rho) = J$?

To answer these two questions we need some further notions about monoids of Lie type. The reader should consult [84] for a detailed account of this theory (notice however that in [84] monoids of Lie type are referred to as *regular split monoids*). Also, the reader needs some familiarity with the representation theory of finite semigroups. For this, the reader is referred to [15] or [119].

We now introduce some notation. This is mainly for convenience, and to reassert the distinction between an idempotent and the $\mathcal{J}$-class it represents. Let Λ be a cross-section lattice and let $e \in \Lambda$. Then $\{e\} = \Lambda \cap J$ and $C_G^r(e) = P_{\lambda(J)}$, where $\lambda(J) \subseteq S$. If $B \subseteq C_G^r(e)$ is a Borel subgroup and $H = \{g \in G | ge = e\}$ then BH is a parabolic subgroup containing B.

Definition 10.11.

$$\nu(J) \in 2^S \textit{ via } BH = P_{\nu(J)}.$$

Notice that $\nu(J) = \lambda_*(e)$, where $\lambda_*(e)$ is as defined in Definition 7.11. Notice also that, if we let $K_J = \{g \in G | ge = eg = e\}$, then $P_{\nu(J)} = BK_J$. We can now state the main theorem (in particular, answering questions a) and b) above).

Theorem 10.12. *Let $I \in 2^S$ and $J \in \mathfrak{U}(M)$. Assume that $\nu(J) \subseteq I \subseteq \lambda(J)$. Define*

$$\alpha_{I,J} = \{\chi : L_I \to \overline{\mathbb{F}}_q^* | \chi(g) = \chi(h) \text{ if } e_J g = e_J h\}.$$

Then there is a one-to-one correspondence between the irreducible representations of M and the set

$$\bigsqcup_{\substack{I \in 2^S, J \in \mathfrak{U}(M) \\ \nu(J) \subseteq I \subseteq \lambda(J)}} \alpha_{I,J}.$$

Under this correspondence $\chi \in \alpha_{I,J}$ corresponds to the unique irreducible representation $\rho : M \to End(M)$ such that

(i) $Apex(\rho) = J$
(ii) there is a line $Y \subseteq V$ such that $\{g \in G | \rho(g)Y = Y\} = P_I$
(iii) if $g \in P_I$ and $y \in Y$ then $\rho(g)(y) = \chi(g)y$.

Proof. Let $\rho : M \to End(V)$ be irreducible with apex $J \in \mathfrak{U}(M)$. Let $J^0 = J \cup \{0\}$, with multiplication defined by

$$xy = \begin{cases} 0 & \text{, if } x = 0,\ y = 0 \text{ or } xy \notin J \\ xy & \text{, if } xy \in J \ . \end{cases}$$

J^0 is a completely 0-simple semigroup (see Definition 2.67) So by Munn-Ponizovskii [15], V is also an irreducible J^0-module; and by Theorem 10.10 above, for $e \in E(J)$, $e(V)$ is an irreducible $H(e)$-module, where $H(e)$ is the unit group of eMe. But $H(e) = eC_G(e)$, and so $e(V)$ is also an irreducible $C_G(e)$-module. Now $C_G(e) \subseteq G$ is the Levi factor of $P_{\lambda(J)} = C_G^r(e)$, and so it is also a finite group of Lie type. Hence Richen's theory applies to the $C_G(e)$-module $e(V)$. Thus, for any Borel subgroup $B_0 \subseteq C_G(e)$ there exists a unique line $Y \subseteq e(V)$ such that $\rho(B_0)Y = Y$.

Let $H = \{g \in G | \rho(g)Y = Y\}$. One checks, as in [114], that $H \subseteq C_G^r(e)$ and H contains a Borel subgroup of G.

Observe that $K_J \subseteq H$, and so $P_{\nu(J)} = BK_J \subseteq H \subseteq P_{\lambda(J)}$. So we can now summarize the relevant properties of an irreducible representation $\rho : M \to End(V)$ with apex $J \in \mathfrak{U}(M)$.

(a) Let $H = \{g \in G | \rho(g)Y = Y\}$. Then $H = P_I$ is parabolic and $\nu(J) \subseteq I \subseteq \lambda(J)$.
(b) If $g \in K_J$ then $\rho(g)y = y$ for all $y \in Y$.
(c) $\rho|G$ is the irreducible representation of type (I, χ), where χ is defined via $\rho(g)(y) = \chi(g)y$ for $g \in L_I$.

On the other hand, suppose that $\rho' : M \to End(V')$ is an irreducible representation with apex J and parabolic subgroup $H' = P_I$ with character χ. Then by Richen's results, $(\rho'|C_G(e), \rho'(e)(V)) \cong (\rho|C_G(e), \rho(e)(V))$ since they have the same (I, χ). But then by Munn-Ponizovskii, ρ and ρ' are equivalent because they come from the same irreducible representation of eMe. Thus, the correspondence

$$\rho \rightsquigarrow (I, J, \chi)$$

is injective. To complete the proof, it remains only to be shown that all possible invariants (I, J, χ) actually arise from irreducible representations of M. But this is now a counting problem. It is easy to check, using Richen's results, that the number of irreducible representations of $H(e)$ is

$$\sum_{\nu(J) \subseteq I \subseteq \lambda(J)} |\alpha_{I,J}|.$$

Thus, by Munn-Ponizovskii, there are exactly

$$\sum_{J \in \mathfrak{U}(M)} \; \sum_{\nu(J) \subseteq I \subseteq \lambda(J)} |\alpha_{I,J}|$$

irreducible representations of M. Hence the above correspondence must be surjective. This concludes the proof.

Let $x \in M_r$. We say that x is *semisimple* if

i) x is a unit in the monoid eM_re for some idempotent e of M_r
ii) $x^k = e$ for some k with $(k, q) = 1$.

Let $M_r^{ss} = \{x \in M_r^{ss} \mid x \text{ is semisimple}\}$ and let $M_r^{ss}/\sim$ denote the set of conjugacy classes of semisimple elements of M_r.

Lemma 10.13. $|IR(M_r)| = |M_r^{ss}/\sim|$.

Proof. By the theory of Munn and Ponizovskii [15],

$$|IR(M_r)| = \sum_{e \in \Lambda} |IR(H(e))|,$$

where $H_r(e)$ is the unit group of eM_re. But from Theorem 42 of [124], $|IR(H_r(e))| = H_r(e)^{ss}/\sim$. Hence it suffices to see that two elements of $H_r(e)$ are $H_r(e)$-conjugate if and only if they are G_r-conjugate. But this is straightforward.

Let $\pi : M \to X$ be the adjoint quotient, as in Corollary 9.4. We need some further assumptions to relate $M_r^{ss}/\sim$ with $X(\mathbb{F}_{q^r})$.

Recall that a reductive monoid M is **locally simply connected** if, for each $e \in E(M)$, H_e has trivial divisor class group. See Definition 6.4.

We say that M is **split** over $\mathbb{F}_q$ if its unit group G is split over $\mathbb{F}_q$ in the usual sense [133]. Recall that any M, defined over $\mathbb{F}_q$, is split over $\mathbb{F}_{q^r}$ for some $r > 0$.

Proposition 10.14. *Let M be a lsc, reductive monoid defined over $\mathbb{F}_q$.*

a) The canonical map $M_r^{ss}/\sim\longrightarrow X(\mathbb{F}_{q^r})$ is bijective for each $r > 0$.
b) $|IR(M_r)| = \sum_{e\in\lambda} q^{r(e)}|H(e)_{ab}(\mathbb{F}_{q^r})|$ where $r(e)$ is the semisimple rank of $H(e)$ and $H(e)_{ab}$ is the abelianization of $H(e)$.
c) If M is split over $\mathbb{F}_q$ then $|H(e)_{ab}(\mathbb{F}_{q^r})| = (q^r - 1)^a$ for some $a \geq 0$.

Proof. From Corollary 9.4, X parametrizes G-conjugacy classes of semisimple elements of M. But now we can apply Theorem 10.3 of [136]. This says that, for each $e \in \Lambda$, $(H^{ss}(\mathbb{F}_q)/\sim) \longrightarrow X(\mathbb{F}_q)$ is bijective. Combined with Theorem 10.10 we obtain our result.

To prove b) one needs a careful calculation combining a) above, Richen's theory [18], Munn-Ponizovskii theory [15], and some basic results from [131]. See Theorem 4.2 of [114] for more details.

For c), first notice that $|H(e)_{ab}(\mathbb{F}_q r)|$ is a factor of $det(\sigma^* - 1)$ using 6.1(d) of [110]. But M is split so that $det(\sigma^* - 1) = (q^r - 1)^m$, where m is the rank of G.

We can now obtain very precise information relating $\{IR(M_r)\}$ and X, for lsc monoids.

First, we recall a key definition (see Definition 7.9).

A reductive monoid M is $\mathfrak{J}$-irreducible if $\mathfrak{U}(M)$ contains exactly one, minimal, nonzero $\mathfrak{J}$-class.

Theorem 10.15. *Let M be lsc and split over $\mathbb{F}_q$ with adjoint quotient $\pi : M \longrightarrow X$. Then*

a) $|IR(M_r)| = 1 = (q^r - 1)\sum_{i\geq 0} b_i q^{ri}$ for some integers b_i independent of $r > 0$;
b) if M is $\mathfrak{J}$-irreducible then

$$|IR(M_r)| - 1 = (q^r - 1)\sum_{i\geq 0} a_i(q^r - 1)^{ri}$$

where

$$a_i = \left|\left\{(I,e) \in 2^S \times \Lambda \middle| \begin{matrix}\nu(e) \subseteq I \subseteq \lambda(e) \\ |\lambda(e)\backslash I| = i\end{matrix}\right\}\right|;$$

c) If M is $\mathfrak{J}$-irreducible then $\mathbb{P}(X) := (X\backslash 0)/K^* = \bigsqcup_{e\in\Lambda\backslash\{0\}} C_e$ where $C_e \cong K^{b_i}$. In particular, b_i is the $2i$-th Betti number of $\mathbb{P}(X)$.

Proof. For a) use Proposition 10.14 b) and c). For b) one needs Theorem 3.1 of [114] which calculates $|IR(M)|$ in terms of $\{(I,e) \in 2^s \times \Lambda \mid \nu(e) \subseteq I \subseteq \lambda(e)\}$. For c) notice that $X = \bigsqcup_{e\in\Lambda} H^{ss}(e)/\sim$. But from Theorem 1.6 of [110], $(H^{ss}(e)/\sim) = K^{b_e} \times K^*$. Now apply Theorem 2.59.

Notice that the cell decomposition in c) above can not be obtained by the method of Birula-Bialynicki [4] and so one must use Theorem 2.59 to obtain these Betti numbers.

Example 10.16. Let $M = M_n(K)$ where $K = \overline{\mathbb{F}_q}$. So $M_r = M_n(\mathbb{F}_{q^r})$. Then, by Munn-Ponizovskii [15],

$$|IR(M_r)| = \sum_{m=0}^{n} |IR(G\ell_m(\mathbb{F}_{q^r}))|$$

while, by Richen [18],

$$|IR(G\ell_m(\mathbb{F}_{q^r}))| = (q^r - 1)q^{r(m-1)} \ .$$

Thus

$$|IR(M_r)| = (q^r - 1)\sum_{i=0}^{n-1} q^{ri} \ .$$

Hence

$$b_i = \begin{cases} 1 \ , \ i = 0, \ldots, n-1 \\ 0 \ , \ \text{otherwise} \ . \end{cases}$$

Plainly, b_i is the 2i-th Betti number of $\mathbb{P}(X) = \mathbb{P}^{n-1}$.

But we can also write

$$|IR(G\ell_m(\mathbb{F}_{q^r}))| = \sum_{i=0}^{m-1} \binom{m-1}{i}(q^r - 1)^{i+1} \ .$$

Hence

$$\begin{aligned} |IR(M_r)| - 1 &= \sum_{m=1}^{n} |IR(G\ell_m(\mathbb{F}_{q^r}))| \\ &= \sum_{i=0}^{n} \binom{n}{i+1}(q^r - 1)^{i+1} \ . \end{aligned}$$

By 6.4b) we obtain the curious combinatorial formula

$$\binom{n}{i+1} = \left| \left\{ (I,e) \in 2^S \times \Lambda \ \middle| \ \begin{matrix} \nu(e) \subseteq I \subseteq \lambda(e) \\ |\lambda(e) \backslash I| = i \end{matrix} \right\} \right| \ .$$

Example 10.17. In [96] the author and M. Putcha construct, for each group G of Lie type, a certain *canonical monoid* $M(G)$ having the following properties.

(a) G is the unit group of $M(G)$.
(b) The type map $\lambda : \mathcal{U}(M(G)) \to 2^S$ of $M(G)$ satisfies
 (i) $\lambda : \mathcal{U}(M(G)) \backslash \{0\} \to 2^S$ is bijective
 (ii) $\lambda(J_e \wedge J_f) = \lambda(J_e) \cap \lambda(J_f)$, where $\mathcal{U}(M(G))$ has been identified with a cross-section lattice Λ of $M(G)$.
(c) For each $e \in \Lambda$, $\{g \in G | ge = eg = e\} = \{1\}$.

By the results of [84], $M(G)$ is determined up to isomorphism by these properties. This monoid also enjoys a number of other useful properties that were important in the proof of Theorem 10.10. In any case, if $J \in \mathfrak{U}(M(G))$ then by (c), $\nu(J) = \phi$. Furthermore, $\alpha_{I,J} = Hom(L_I, \overline{\mathbb{F}}_q^*)$ for any $I \subseteq \lambda(J)$. So $\alpha_{I,J}$ is independent of J if it is non-empty.

Define, for any finite monoid M of Lie type,

$$IR(M) = \{\rho : M \to End(V) | \rho \text{ is irreducible}\}/ \sim$$

where "$\sim$" denotes equivalence of representations. Thus, by Theorem 10.10,

$$\begin{aligned} |IR(M(G))| &= \sum_{I \subseteq \lambda(J)} |\alpha_{I,J}| \\ &= \sum_{I \subseteq S} \sum_{J \in \mathfrak{U}(S)} |\alpha_{I,J}| \\ &= \sum_{I \subseteq S} 2^{|S \setminus I|} \alpha_I \end{aligned}$$

where α_I is the common value of $|\alpha_{I,J}|$ for $I \subseteq \lambda(J)$. This agrees with the formula (1) in the proof of Theorem 2.2 of [96].

If $G = S\ell_{n+1}(\mathbb{F}_q)$ then $|S| = n$ and, for $|I| = i, \alpha(I) = (q-1)^{n-i}$. Thus

$$\begin{aligned} |IR(M(G))| &= \sum_{I \subseteq S} 2^{|S \setminus I|} \alpha(I) \\ &= \sum_{i=0}^{n} \binom{n}{i} 2^{n-i} (q-1)^{n-i} \\ &= \sum_{i=0}^{n} \binom{n}{i} (2q-2)^{n-i} \\ &= (2q-1)^n. \end{aligned}$$

Notice that this corrects the calculation error of Example 2.3 of [96].

The reader might wonder if, for any reductive monoid M, there is a finite, dominant morphism

$$\varphi : M' \to M$$

of reductive monoids with M' a lsc monoid. In general this seems to be a delicate problem. However, we do have some positive results.

If M is reductive and normal we denote by $Cl(M)$ the *divisor class group* of M. See § 2.1.3 for a summary of some of the main properties relevant to our discussion.

Recall the following results from Theorem 6.7.

Theorem 10.18. *Let M be reductive and normal.*

a) If $Cl(M) = (0)$ then M is lsc.
b) If $M \backslash G$ is irreducible then $Cl(M)$ is finite and there exists $\pi : M' \longrightarrow M$, finite and dominant, such that $Cl(M') = (0)$.

If M is a $\mathfrak{J}$-coirreducible monoid then the lattice of $\mathfrak{J}$-classes $\Lambda \cong \mathfrak{U}(M)$, and the type map $\lambda : \Lambda \to 2^S$ are both determined by $\lambda(J)$ where $J \in \mathfrak{U}(M) \backslash \{1\}$ is the unique maximal element. Indeed, if $\lambda(J) = I$, then

$$\Lambda = \left\{ X \in 2^S \middle| \begin{array}{l} \text{no component of } X \text{ is} \\ \text{contained in } I \end{array} \right\} \cup \{1\}$$

where $\Lambda \backslash \{1\}$ is ordered by reverse inclusion and $1 \in \Lambda$ is the largest element. Here, X corresponds to $e_X \in \Lambda$. Furthermore, $\lambda : \Lambda \to 2^S$ is defined by

$$\lambda(X) = X \sqcup C_I(X)$$

if $X \neq \phi$ or 1, and $\lambda(\phi) = I$ and $\lambda(1) = S$. Here, $C_I(X) = \{\alpha \in I | \sigma_\alpha \sigma_\beta = \sigma_\beta \sigma_\alpha$ for all $\beta \in X\}$. It follows from Definition 7.11 and Lemma 7.15 that $\lambda_*(e_X) = X$ and $\lambda^*(e_X) = C_I(X)$.

Theorem 10.19. *Suppose that M is $\mathfrak{J}$-coirreducible of type I, split over $\mathbb{F}_q$ and with $Cl(M) = 0$. Then*

$$|IR(M)| = \sum_{X \in \Lambda} q^{|C_I(X)|}(q-1)^{|S \backslash \lambda(X)|} + q^{|S|}(q-1).$$

Proof. First notice that $|S \backslash X| - |C_I(X)| = |S \backslash \lambda(X)|$ since $\lambda(X) = X \cup C_I(X)$ is a disjoint union whenever no component of X lies in I. But then for each $X \in \Lambda$, $IR(H(e_x)) = q^{|C_I(X)|}(q-1)^{|S \backslash \lambda(X)|}$ since $H(e_x)$ has rank $|S \backslash X|$ and semisimple rank $|C_I(X)|$.

Theorem 10.20. *Suppose that $\underline{M}$ is reductive, split over $\mathbb{F}_q$ and locally simply connected. Then*

$$|IR(M)| = \sum_{i=0}^{|S|+1} a_i (q-1)^i$$

where $a_i = |\{(I, J) \in 2^S \times \mathfrak{U}(M) | rank(J) - |I| + |v(J)| = i\}|$.

Proof. Recalling Theorem 10.12 it suffices to prove that, given our assumptions on M, $|\alpha_{I,J}| = (q-1)^i$ where $i = rank(J) - |I| + |\nu(J)|$. But $\alpha_{I,J}$ is the character group of $L_{I \backslash \nu(J)}(H(e))$, the Levi subgroup of H_e of type $I \backslash \nu(J)$. This has rank equal to rank(J) and semisimple rank $|I| - |\nu(J)|$. This yields the formula for a_i.

10.6 Exercises

10.6.1 Weil Zeta Functions

Let X be a variety defined over the finite field $\mathbb{F}_q$, and let

$$Z(X) = exp(\sum_{r\geq 1} \frac{N_r}{r} t^r)$$

be its Weil zeta function.

1. Suppose there exist complex numbers $\alpha_1, \ldots, \alpha_s$ and $\beta_1, \ldots, \beta_t$ such that, for all $m \geq 1$,

$$N_m = \sum_{i=1}^{s} \alpha_i^m - \sum_{j=1}^{t} \beta_j^m.$$

Show that

$$exp(\sum_{r\geq 1} \frac{N_m}{m} t^m) = \frac{\Pi_{j=1}^t (1-\beta_j t)}{\Pi_{i=1}^s (1-\alpha_i t)}.$$

2. Show that if $Z(X) = \Pi_{j\geq 0}(1-q^j t)^{b_j}$, then $Z(X \times K^n) = \Pi_{j\geq 0}(1-q^{j+n}t)^{b_j}$.
3. Show that

$$Z(X \times K^*) = \frac{Z(X \times K)}{Z(X)}.$$

4. Letting $Z(X) = Z(t)$, show that

$$Z(X \times (K^*)^n) = \Pi_{s=0}^n Z(q^s t)^{(-1)^{n-s}\binom{n}{s}}.$$

5. Recall that, if M is reductive and split over $k = \mathbb{F}_q$, then $M = \sqcup_{r\in R} BrB$ in such a way that, if $BrB \cong (K^*)^r \times K^l$, then $(BrB)(k) \cong (k^*)^r \times k^l$, over $k = \mathbb{F}_q$. Show that

$$Z(M) = \Pi_{k=0}^m (1-q^k t)^{b_k}$$

where $m = dim(M) - rank(M)$, and

$$b_k = \sum_{r=0}^{N} \sum_{l=k-r}^{k} (-1)^{r-k+l} \binom{r}{k-l} \nu_{r,l}$$

with

$$\nu_{r,l} = |\{r \in R \mid BrB \cong (k^*)^r \times k^l\}|.$$

10.6.2 Counting Modular Representations

1. Let M be $\mathfrak{J}$-irreducible, split over $\mathbb{F}_q$, and locally simply connected. Write

$$|IR(M(\mathbb{F}_q))| - 1 = (q-1)\sum_{i\geq 0} a_i(q-1)^i$$
$$= (q-1)\sum_{i\geq 0} b_i q^i.$$

   a) Show that $a_i = |\{(I,e) \in 2^S \times \Lambda \mid \lambda_*(e) \subseteq I \subseteq \lambda(e)\}|$.
   b) Show that $b_i = |\Lambda_i|$, the number of $G \times G$-orbits of rank i.
   c) Show that $a_i = b_i + (i+1)b_{i+1} + \cdots + \binom{n}{i} b_n$.

11

Cellular Decomposition of Algebraic Monoids

The most commonly studied cell decompositions in algebraic geometry are those of Bialynicki-Birula [4]. If $S = K^*$ acts on a smooth complete variety X with finite fixed point set $F \subseteq X$, then $X = \bigsqcup_{\alpha \in F} X_\alpha$ where $X_\alpha = \{x \in X \mid \lim_{t \to 0} tx = \alpha\}$. Furthermore, X_α is isomorphic to an affine space. We refer to X_α as a *BB-cell.* If, further, a reductive group G acts on X extending the action of S, we may assume (replacing S if necessary) that each X_α is stable under the action of some Borel subgroup B of G with $S \subseteq B$. In case X is a complete homogeneous space for G, each cell X_α turns out to consist of exactly one B-orbit.

Let M be a *semisimple* monoid. That is, M is reductive, normal with zero element and with one-dimensional center. Define

$$X = (M \backslash \{0\})/K^* \,.$$

Then X is projective and $G \times G$ acts on X by the rule $(g, h) \cdot [x] = [gxh^{-1}]$. Furthermore, any generic one-parameter subgroup $S = K^* \subseteq G \times G$ has a finite number of fixed points on X.

Each BB-cell on X is made up of a finite number of $B \times B$-orbits. But there is often no explicit algorithm for deciding how each BB-cell is made up from the $B \times B$-orbits. On the other hand, we have explicitly identified these $B \times B$-orbits on M (or X) in Chapter 8. So what we need here is a more direct definition, guided by the BB-procedure, that simply tells us how each cell is made up from $B \times B$-orbits.

In this chapter, we define a notion that yields a decomposition of $M\backslash\{0\}$ into a disjoint union of "monoids cells". These cells are defined directly in terms of Λ, B and R. We then explain how to use these monoid cells to obtain an explicit decomposition of the "wonderful compactification" into a disjoint union of affine space.

In the case where X is the wonderful compactification of the adjoint, semisimple group G, this "BB-procedure" has been carried out in [21]. In fact, they obtain results for a more general class of wonderful compactifications. Let G be a semisimple algebraic group, and suppose that $\sigma : G \longrightarrow G$ is an involution (so that $\sigma \circ \sigma = id_G$) with $H = \{x \in G \mid \sigma(x) = x\}$. The *wonderful compactification* of G/H (according to [21]) is the unique normal G-equivariant compactification X of G/H obtained by considering an irreducible representation $\rho : G \longrightarrow G\ell(V)$ of G with $\dim(V^H) = 1$ and with highest weight in general position. Then let $h \in V^H$ be nonzero and define

$$X = \overline{\rho(G)[h]} \subseteq \mathbb{P}(V),$$

the Zariski closure of the orbit of $[h]$. (See Section 2 of [21] for details.) In this chapter we restrict our attention to the special case where the group is $G \times G$ and $\sigma : G \times G \longrightarrow G \times G$ is given by $\sigma(g, h) = (h, g)$. It is easy to see that, in this case, the $G \times G$-variety $(G \times G)/H$ can be canonically identified with G with its two-sided G-action.

Much important work has been accomplished since [21] appeared. In particular, Brion [12] obtains much information about the structure of X. Among other things, he finds a BB-decomposition $X = \bigsqcup_{x \in F} C_x$ from which he then obtains a basis of the Chow ring of X of the form $\{\overline{By(x)B}\}_{x \in F}$. He also identifies explicitly how each cell C_x is made up from $B \times B$-orbits. Since then, Springer [130] and Kato [46] have uncovered more geometry related to this problem.

It appears that my cell decomposition agrees with the one of [12] (although we have not actually verified this). The interested reader should consult Brion's paper, as well as those of Kato and Springer, for more information on the $B \times B$-orbit closures and the Chow ring for X.

11.1 Monoid Cells

In this section we assume that M is a $\mathcal{J}$-irreducible, reductive monoid as in § 7.3. Let $B \subseteq G$ be a Borel subgroup with maximal torus $T \subseteq B$. Define

$$R_1 = \{x \in R \mid xT = Tx \text{ is one-dimensional}\},$$

the set of *rank one elements* of the Renner monoid. Our cells are canonically indexed by R_1.

Let $r \in R_1$. Then there exist unique rank one idempotents $e, f \in E_1(\overline{T})$ such that

$$r = erf.$$

We define the **monoid cell** C_r as follows:

$$C_r = \{y \in M \mid eBy = eBey \subseteq rB\}.$$

The following results are easily obtained.

a) $M\backslash\{0\} = \bigsqcup_{r\in R_1} C_r$.

b) Any BB-decomposition of $(M\backslash 0)/K^*$ coming from a $1-PSG$ $K^* \subseteq G\times G$ with finite fixed point set has exactly $|R_1|$ BB-cells.

So it is likely that these cells could be obtained from the BB-decomposition of some regular $1 - PSG$ $K^* \subseteq G \times G$.

Example 11.1. Let $M = M_n(K)$ with B and T as usual. Then

$$R_1 = \{r_{ij} \mid 1 \leq i,\ j \leq n\}$$

where r_{ij} is the elementary matrix (a_{st}) with $a_{ij} = 1$ and $a_{st} = 0$ for $(s,t) \neq (i,j)$. Then

$$\begin{aligned} C_{r_{ij}} &= \left\{ (a_{pq}) \in M_n(K) \left| \begin{matrix} a_{ij} \neq 0 & & & \\ a_{pq} = 0 & \text{if} & p > i & \\ a_{pq} = 0 & \text{if} & p = i & \text{and} \quad q < i \end{matrix} \right. \right\} \\ &\cong K^{n(i-1)+(n-j)} \times K^* \,. \end{aligned}$$

Recall from Theorem 7.12, that the structure of any $\mathfrak{J}$-irreducible monoid M is largely determined by $J_0 \subseteq S$. Here

$$J_0 = \{s \in S \mid se = es (= e)\},$$

where $e \in \Lambda\backslash\{0\}$ is the unique minimal element of $\Lambda\backslash\{0\}$. We say that M is a *canonical monoid* if $J_0 = \phi$. By 7.12 a),

$$\mu : \Lambda\backslash\{0\} \longrightarrow \{I \subseteq S\}$$

is an order-preserving bijection for any canonical monoid M. For $I \subseteq S$ we write $e_I = \mu^{-1}(I)$.

Proposition 11.2. *a) Let $x \in R$. Then there exist unique $u, v \in W$ and $e_I \in \Lambda$ such that*

i) $x = ue_I v$, and

ii) $I \subseteq \{s \in S \mid \ell(us) > \ell(u)\} := I_u$.

b) For $r = ue_\phi v \in R_1$ define

$$\mathfrak{C}_r = \left\{ x = ue_I v \in R \left| \begin{matrix} u, e_I \text{ and } v \text{ as in a)} \\ I \subseteq I_u \end{matrix} \right. \right\} .$$

Then $C_r = \bigsqcup_{x\in\mathfrak{C}_r} BxB$.

Proof. For a) we write $x = we_I y$ for some $I \subseteq S$ and $w, y \in W$. By Theorem 4.5 c), $\mathfrak{C}_I$ is unique. But from well-known results from Coxeter groups we can write $w = uc$ where $c \in W_I$ and $\ell(us) > \ell(u)$ for any $s \in I$. But $e_I c = ce_I$. So we write $x = ue_I v$ where $v = cy$.

We leave part b) to the reader, since it is not really needed in this survey. Indeed, we can use $\mathfrak{C}_r$ to define C_r.

We refer to $x = ue_I v$ as the *normal form* for x.

We now determine the structure of each C_r. But first we determine how the $B \times B$-orbits fit together.

Proposition 11.3. *Let $x \in R$ and write $x = ue_I v$ in normal form. Then*

$$\begin{aligned} BxB &= (Uu \cap uU)(e_I T)(vU \cap U^- v) \\ &\cong (Uu \cap uU) \times e_I T \times (vU \cap U^- u) \end{aligned}$$

Proof. Let $e = e_I$. Since $eB \subsetneqq Be$ we get $eB = C_B(e)$. Also we have $vBv^{-1} = (vBv^{-1} \cap B)(vBv^{-1} \cap U^-)$ (direct product). Then

$$\begin{aligned} evBv^{-1} &= e(vBv^{-1} \cap B)(vBv^{-1} \cap U^-) \\ &= Ve(vBv^{-1} \cap U^-) \end{aligned}$$

where $V \subseteq (C_b(e)$ is some connected subgroup with $T \subseteq V$. Then we get

$$evB = Ve(vB \cap U^- v). \tag{$*$}$$

We now look at Bue. Recall first that $\ell(us) < \ell(u)$ for any $s \in I$. This is the same as saying that $C_B(e) \subseteq u^{-1}Bu \cap B$. Thus

$$u^{-1}Bue = (u^{-1}Bu \cap B)e$$

since $(u^{-1}Bu \cap U^-)e = \{e\}$. Thus

$$Bue = (Bu \cap uB)e \tag{$**$}$$

Combining (*) and (**) we obtain that

$$\begin{aligned} u^{-1}BuevB &= (u^{-1}Bu \cap B)e(VvB \cap U^- v) \\ &= (u^{-1}Bu \cap B)e(vB \cap U^- v) \end{aligned}$$

since $V \subseteq C_B(e) \subseteq u^{-1}Bu \cap B$. Thus

$$BuevB = (Uu \cap uU)(eT)(vU \cap U^- v)\ .$$

A calculation similar to the proof of Lemma 5.1 shows that this product is direct. In characteristic $p > 0$, the product morphism is seen to be separable by a local analysis of the torus action at the fixed point.

Proposition 11.3 tells us exactly why we have found the "correct" definiton of cells.

Proposition 11.4. *Let $r = ue_\phi v \in R_1$.*

a) $C_r \cong (Uu \cap uU) \times Z_u \times (vU \cap U^- v)$ where $Z_u = \bigsqcup_{J \subseteq I_u} e_I T$.

b) $Z_u/K^* \cong K^{i(u)}$ *where* $i(u) = |I_u|$.

Proof. For b) we may assume that $u = 1$, since any other Z_u is a T-orbit closure in $Z_1 = \bigsqcup_{I \subseteq S} e_I T$. But one checks, as in Proposition 3.4 of [118], that $\mathcal{O}(Z_1/K^*) = K[\alpha_1^{-1}, \dots, \alpha_s^{-1}]$ where $S = \{\alpha_1, \dots, \alpha_s\}$.

For a), we have by definition

$$\begin{aligned} C_r &= \bigsqcup_{ue_Iv \in \mathcal{C}_r} (Uu \cap uU)(e_I T)(vU \cap U^- v) \\ &= (Uu \cap uU)\left(\bigsqcup_{ue_Iv \in \mathcal{C}_r} e_I T\right)(vU \cap U^- v) . \end{aligned}$$

Theorem 11.5. *Let* $r = ue_\phi v \in R_1$. *Then there is a bijective morphism*

$$m : K^{n_r} \longrightarrow C_r/K^*$$

where $n_r = \ell(w_0) - \ell(u) + i(u) + \ell(\sigma)$. *Here,* $w_0 \in W$ *is the longest element, so that* $\ell(w_0) = |\Phi^+|$.

Proof. From 11.4 a), dim $C_r = \dim(Uu \cap uU) + \dim(Z_u) + \dim(vU \cap U^- v)$. One checks that $\dim(Uu \cap uU) = \ell(w_0) - \ell(u)$ and that $\dim(vU \cap U^- v) = \ell(v)$. From 11.4 b) $\dim(Z_u) = i(u) + 1$.

We can also consider the cell decomposition for each orbit closure. Let $I \subseteq S$ and define

$$X_I = (\overline{Ge_I G} \backslash \{0\})\big/K^* .$$

By [21] X_I is a smooth spherical $G \times G$-subvariety of X. We find a cell decomposition of X_I as follows.

Given $r \in R_1$ and $I \subseteq S$ define

$$C_{I,r} = C_r \cap X_I .$$

Clearly $X_I = \bigsqcup_{r \in R_1} C_{I,r}$. But we can say more.

Theorem 11.6. *Let* $r = ue_\phi v \in R_1$. *Then there is an isomorphism*

$$m : K^{n_{I,r}} \longrightarrow C_{I,r}$$

where $n_{I,r} = \ell(w_0) - \ell(u) + |I \cap I_u| + \ell(v)$.

Proof. By inspection $C_{I,r} = \bigsqcup_{j \subseteq I \cap I_u} Bue_J vB$, and so

$$C_{I,r} = (Uu \cap uU)(Z_{I,u})(vU \cap U^- v)$$

where $Z_{I,u} = \bigsqcup_{J \subseteq I \cap I_n} e_J T$. Then the proof proceeds as in 11.4 and 11.5.

If Y is a smooth projective algebraic variety with cell decomposition we can calculate the Betti numbers of Y:

$$\beta_{2i}(Y) = \text{ the number of cells of (complex) dimension } i.$$

Let

$$P(X,t) = \sum_{i \geq 0} \beta_{2i}(X) t^{2i}$$

be the Poincaré polynomial of X.

Theorem 11.7.

$$P(X_I, t) = \left(\sum_{u \in W} t^{2(\ell(w_0) - \ell(u) + |I_u \cap I|)} \right) \left(\sum_{v \in W} t^{2\ell(v)} \right) .$$

This result is also obtained by DeConcini and Procesi in [21].

It appears that Springer has obtained the same cell decomposition in [130], in his detailed study of the geometry of the $B \times B$-orbit closures of X.

Further results about X have been obtained by Kato [46]. In particular, he obtains a kind of Borel-Weil theorem for X.

Since the appearance of [21], these same authors have continued the study of these interesting spaces. In [22] they have described the rational cohomology of X, as well the cohomology of many other spaces closely related to X. In [5] (along with Bifet), they describe the rational cohomology ring of any complete symmetric variety by generators and relators. See § 15.4 for a brief description of these developments.

11.2 Exercises

Let W be a Weyl group with generating set $S \subseteq W$, the set of simple involutions, and let $\Delta \subseteq \Phi$ be the set of positive, simple roots.

1. For $x \in W$, define $I_x = \{\alpha \in \Delta \mid l(s_\alpha x) = l(x) - 1\}$ and $J_x = \{\alpha \in \Delta \mid l(x s_\alpha) = l(x) - 1\}$. Show that, for all $x \in W$, $I_x = J_{x^{-1}}$.
2. For a subset $\theta \subseteq S$ let $w(\theta)$ be the longest element of Coxeter group W_θ. If $x \in W$, let $w_x = w(I_x)$. Show that, for all $x \in W$, $x = w_x x_-$, where $l(x) = l(w_x) + l(x_-)$.
3. For $s \in S, y \in W$, define

$$s * y = \begin{cases} y & \text{, if } l(sy) < l(y) \\ sy & \text{, if } l(sy) > l(y) . \end{cases}$$

Show that $*$ defines on W the structure of a monoid with the following properties:

a) $w(\theta) * x = x$ if and only if $\theta \subseteq I_x$

b) $x * v = v$ if and only if $x * w_v = w_v * x = w_v$.

4. Let $x \in W$ and $s \in S$. Prove the following:
 a) If $l(xs) = l(x) + 1$ then $I_{xs} = I_x \cup (\{xsx^{-1}\} \cap S)$.
 b) If $l(xs) = l(x) - 1$ then $I_x = I_{xs} \cup (\{xsx^{-1}\} \cap S)$.
 c) $I_{xs} = I_x$ if and only if $xsx^{-1} \notin S$.
5. Let $x \in W$. Prove that

$$|\{x \in W \mid I_x = I\}| = \sum_{I \subseteq J} (-1)^{|J \setminus I|} \frac{|W|}{|W_J|}.$$

6. Show that, for $I \subseteq S$,

$$\sum_{x \in W,\, I_x = I} t^{l(x)} = \sum_{I \subseteq J} (-1)^{|J \setminus I|} t^{dim(G/P_J)} P(J,t)$$

where $P(J,t) = \sum_{j \geq 0} dim(H^{2j}(G/P_J))t^j$ is the Poincaré polynomial of G/P_J.

12

Conjugacy Classes

In this chapter we describe Putcha's theory of conjugacy classes in a reductive monoid. We can not do justice here to this truly remarkable development. So we shall refer the reader to Putcha's work [88, 89, 90] for the details of many proofs. Our purpose here is to explain Putcha's main results while describing some of the key ideas of his proofs.

The basic idea here is to define, for each $(e, \sigma) \in E(\overline{T}) \times W$, a subset $M_{e,\sigma} \subseteq M$ such that

a) any $x \in M$ is conjugate to some $y \in M_{e,\sigma}$ for some $(e, \sigma) \in E(\overline{T}) \times W$
b) there is an explicit equivalence relation $\sim$ on $M_{e,\sigma}$ such that for $a, b \in M_{e,\sigma}$, $a \sim b$ if and only if $b = gag^{-1}$ for some $g \in G$.

On this basis we discuss some further refinement, especially the issue of finding a minimal collection $A = \{(e, \sigma)\}$ so that $M = \bigcup_{(e,\sigma)\in A} gM_{e,\sigma}g^{-1}$. The key results here can be described using some finer structure theory of the Renner monoid.

12.1 The Basic Conjugacy Theorem

Let M be a reductive monoid with zero element $0 \in M$ and unit group $G \subseteq M$. Let $e \in E(\overline{T})$ and $\sigma \in W$. For $\theta \in W$ write

$$e^\theta = \theta^{-1} e \theta \in E(\overline{T}) \ .$$

Define

$$M_{e,\sigma} = eC_G(e^\theta \mid \theta \in \langle \sigma \rangle)\sigma \ .$$

Theorem 12.1. *Let $a \in M$. Then a is conjugate to an element of $M_{e,\sigma}$ for some e, σ.*

Proof. By the basic results of reductive monoids (Corollary 2.3 of [84]) there is a maximal torus $T_1 \subseteq G$ and idempotents $e, f \in E(\overline{T_1})$ such that $e\mathcal{R}a\mathcal{L}f$. Hence there exists $\theta = mT \in W$ such that $e^\theta = f$. Hence $em\mathcal{H}a$. Thus $a \in eC_G(e)m = eC_G(e)\theta$. This is the beginning of an induction argument. Inductively, assume that

$$a \in eC_G(e^{\theta^j} \mid j = 0, 1, \ldots, k)\theta$$

where $k \geq 0$. One can then find $b \in eC_G(e^{\theta^j} \mid j = 0, 1, \ldots, k+1)\theta$ such that a is conjugate to b, using Theorem 2.2 of [88].

Now we must determine the exact conditions for two elements of $M_{e,\sigma}$ to be conjugate.

Fix $e \in E(\overline{T})$ and $\sigma = nT \in W$. Let

$$V = C_G(e^\theta \mid \theta \in \langle\sigma\rangle) .$$

Then V is reductive, $T \subseteq V$ and $V^\sigma = V$ (where $X^\sigma := n^{-1}Xn$). Let

$$\begin{aligned} \widetilde{V}_e &= \{a \in V \mid ae = ea = e\} \\ &= \widetilde{T}_e V_e , \end{aligned}$$

where $\widetilde{T} = \{t \in T \mid te = et = e\ \}$, and $V_e \in V$ is the connected component of the identity element of V.

Then $\widetilde{V}_e$ is a closed normal subgroup of V. Finally, let

$$\Omega = \prod_{\theta \in \langle\sigma\rangle} (\widetilde{V}_e)^\theta .$$

Then Ω is a closed, normal subgroup of V.

Definition 12.2. *a) If $x \in V$ let $x^* = nx^{-1}n^{-1} \in V$.*
b) Let $G_{e,\sigma} = V/\Omega$ and define $\zeta : M_{e,\sigma} \longrightarrow G_{e,\sigma}$ as follows: For $a = evn \in M_{e,\sigma}$, $v \in V$, let

$$\zeta(a) = v\Omega \in G_{e,\sigma} .$$

Since $\widetilde{V}_e \subseteq \Omega$, ζ is well-defined.

Theorem 12.3. *Let $a, b \in M_{e,\sigma}$. Then a and b are conjugate in M if and only if there exists $x \in G_{e,\sigma}$ such that $x\zeta(a)x^* = \zeta(b)$.*

Proof. We refer the reader to Theorem 2.4 of [88] for an amazing display.

Example 12.4. If $M = M_n(K)$ then the group $G_{e,\sigma}$ is trivial if $e\sigma$ is nipotent.

Example 12.5. Let $r \geq 2$ and define $J_r = \begin{pmatrix} 0 & & 1 \\ & \cdot^{\cdot^{\cdot}} & \\ 1 & & 0 \end{pmatrix}$. Define for $n \geq 2$,

$$G_0 = \left\{ A \in S\ell_{2n+1}(K) \,\middle|\, A^t \left(\begin{array}{c|c} 1 & \\ \hline & J_{2n} \end{array} \right) A = \left(\begin{array}{c|c} 1 & 0 \\ \hline 0 & J_{2n} \end{array} \right) \right\} .$$

Then G_0 is the special orthogonal group of type B_n. Let $G = G_0 K^*$ and $M = \overline{G} \subseteq M_{2n+1}(K)$. Then

$$e = \begin{pmatrix} 0 & 0 & 0 \\ 0 & I_n & 0 \\ 0 & 0 & 0 \end{pmatrix} , \quad f = \begin{pmatrix} 0 & 0 & 0 \\ 0 & 0 & 0 \\ 0 & 0 & I_n \end{pmatrix} \in E(M) .$$

If $\sigma = \begin{pmatrix} 1 & 0 \\ 0 & J_{2n} \end{pmatrix} \in W$ then $e^\sigma = f$ and $(e\sigma)^2 = 0$. One can check that $G_{e,\sigma} \cong PG\ell_n(K)$ with $* : G_{e,\sigma} \longrightarrow G_{e,\sigma}$ defined by $A \longmapsto J_n A^t J_n$.

Remark 12.6. Three important questions cry out for an answer here.

a) When is $gM_{e,\sigma}g^{-1} \cap M_{f,\tau} \neq \phi$?
b) If $gM_{e,\sigma}g^{-1} \cap M_{f,\tau} \neq \phi$, can we conclude that $\bigcup_{g \in G} gM_{e,\sigma}g^{-1} = \bigcup_{g \in G} gM_{f,\tau}g^{-1}$?
c) If $e\sigma = e\tau$, how are $M_{e,\sigma}$ and $M_{e,\tau}$ related?

12.2 Some Refinements

In this section we describe the results of [89]. We arrive at Putcha's definitive answers to the above three questions. Define

$$N_{e,\sigma} \subseteq M_{e,\sigma}$$

by

$$\begin{aligned} N_{e,\sigma} &= eC_G(T_{e^\theta} \mid \theta \in \langle\sigma\rangle)\sigma \\ &= eC_G(G_{e^\theta} \mid \theta \in \langle\sigma\rangle)T_\sigma \\ &= eC_G(G_{e^\theta} \mid \theta \in \langle\sigma\rangle)\sigma . \end{aligned}$$

Clearly, $N^\pi_{e,\sigma} = N_{e,\sigma\pi}$ for $\pi \in W(e)$. Let $\pi \in W_e$. Then $\pi = mT$ for some $M \in G_e \cap N_G(T)$. Let $a \in N_{e,\sigma}$. Then $a = egn$ for some $g \in C_G\ (G_{e^\theta} \mid \theta \in \langle\sigma\rangle)$, $n \in N_G(T)$ with $\sigma = nT$. So for all $i \geq 0$, $n^i g n^{-i} \in C_G(G_e)$ and so $n^i g n^{-i}$ is centralized by m. By induction on i,

$$(mn)^i g (mn)^{-i} = mn^i g n^{-i} m^{-1} = n^i g n^{-i} \in C_G(G_e) .$$

Hence $g \in C_G(G_{e^\theta} \mid \theta \in \langle\sigma\rangle)$ and so

$$egn = emgn = egmn \in N_{e,\pi\sigma} .$$

We conclude that $N_{e,\sigma} \subseteq N_{n,\pi\sigma}$. Similarly, $N_{e,\pi\sigma} \subseteq N_{e,\sigma}$. So

$$N_{e,\sigma} = N_{e,\pi\sigma} \text{ for all } \pi \in W_e .$$

For this reason we write $N_{e\sigma}$ for $N_{e,\sigma}$. Hence $N_{e\sigma}$ depends only on the element $e\sigma$ of R.

Example 12.7. Let $M = M_5(K)$, and let

$$e = \begin{pmatrix} 1&0&0&0&0\\ 0&1&0&0&0\\ 0&0&0&0&0\\ 0&0&0&0&0\\ 0&0&0&0&0 \end{pmatrix}, \sigma = \begin{pmatrix} 0&0&1&0&0\\ 0&0&0&1&0\\ 1&0&0&0&0\\ 0&1&0&0&0\\ 0&0&0&0&1 \end{pmatrix}, \tau = \begin{pmatrix} 0&0&1&0&0\\ 0&0&0&1&0\\ 0&1&0&0&0\\ 0&0&0&0&1\\ 1&0&0&0&0 \end{pmatrix} .$$

Then

$$M_{e,\sigma} = \left\{ \begin{pmatrix} 0&0&a&b&0\\ 0&0&c&d&0\\ 0&0&0&0&0\\ 0&0&0&0&0\\ 0&0&0&0&0 \end{pmatrix} \;\middle|\; ad \neq bc \right\} \quad \text{yet} \quad M_{e,\tau} = \left\{ \begin{pmatrix} 0&0&a&0&0\\ 0&0&0&b&0\\ 0&0&0&0&0\\ 0&0&0&0&0\\ 0&0&0&0&0 \end{pmatrix} \;\middle|\; ab \neq 0 \right\} .$$

But also $e\sigma = e\tau$ and $M_{e,\tau} = N_{e\tau} = N_{e\sigma}$.

Theorem 12.8. *a) If $r, s \in R$ and $N_r \cap N_s \neq \phi$ then $N_r = N_s$.*
b) If $\delta \in C_W(e^\theta \mid \theta \in \langle\sigma\rangle)$ then $N_{e\delta\sigma} \subseteq M_{e,\sigma}$ and $N_{e\delta\sigma} = N^\pi_{e\sigma}$ for some $\pi \in C_W(e^\theta \mid \theta \in \langle\sigma\rangle)$.
c) Any element of $M_{e,\sigma}$ is conjugate to some element of $N_{e\sigma}$.
d) The map $\zeta : M_{e,\sigma} \longrightarrow G_{e,\sigma}$ remains surjective when restricted to $N_{e\sigma} \subseteq M_{e,\sigma}$.

Proof. See Theorem 2.3 of [89].

Putcha goes on to obtain the following spectacular result.

Theorem 12.9. *The following are equivalent.*

a) Some element of $M_{e,\sigma}$ is conjugate to an element of $M_{e,\theta}$.
b) $\bigcup_{g\in G} gM_{e,\sigma}g^{-1} = \bigcup_{g\in G} gM_{e,\theta}g^{-1}$.
c) There exists $\gamma \in W$, with $e\theta$ conjugate to $e\gamma$ in R, such that

$$\bigcap_{i\geq 0} \gamma^i W(e)\sigma^{-i} \neq \phi .$$

d) $N^{\pi}_{e\sigma} = N_{e\theta}$ *for some* $\pi \in W(e)$.

This is nearly complete. The only thing remaining is to isolate the "best" representative in each set

$$\left\{ e\tau \in R \mid \bigcup_{g \in G} gM_{e,\sigma}g^{-1} = \bigcup_{g \in G} gM_{e,\tau}g^{-1} \right\} .$$

This will be the main focus of the next section.

12.3 Putcha's Decomposition and the Nilpotent Variety

In this section we determine a subset $P \subseteq R$ such that

$$M = \bigsqcup_{r \in P} X(r) \qquad (*)$$

where $X(r) = \bigcup_{g \in G} gM_{e,\sigma}g^{-1}$ for $r = e\sigma \in R$. We shall refer to (*) as the *Putcha decomposition* of M. We also describe the order relation on P that corresponds to the condition $X(r) \subseteq \overline{X(s)}$. This stunning development has no analogue in group theory.

Recall now the Weyl group W and its set of Coxeter group generators $S \subseteq W$. For $I \subseteq S$ define

$$D_I = \{g \in W \mid \ell(gw) = \ell(g) + \ell(w) \text{ for all } w \in W_I\}$$
$$D_I^{-1} = \{g \in W \mid \ell(wg) = \ell(w) + \ell(g) \text{ for all } w \in W_I\} .$$

Definition 12.10. *Let* $x, y \in W$. *We say that* $x \equiv_I y$ *if* $\bigcap_{i \geq 0} x^i W_I y^{-1} \neq \phi$.

Notice that $\equiv_I$ is an equivalence relation on W. Also notice that

$$x \equiv_I wxs^{-1} \text{ if } w \in W_I,$$

and

$$x \equiv_I ux \text{ if } u \in \bigcap_{i \geq 0} x^i W_I x^{-i} .$$

Proposition 12.11. *Let* $x \in W$. *Then* $x \equiv_I y$ *for some unique* $y \in D_I^{-1}$. *Furthermore,* $\ell(y) \leq \ell(x)$.

Proof. Let

$$\begin{aligned} x_1 &= x = w_1 y_1 \quad &, w_1 \in W_I, y_1 \in D_I^{-1} \\ x_2 &= y_1 w_1 = w_2 y_2 &, w_2 \in W_I, y_2 \in D_I^{-1} \\ &\vdots \end{aligned}$$

Now $x_{j+1} = y_j w_j = w_j^{-1}(w_j y_i) w_j = w_j^{-1} x_j w_j$. Hence by comments following 12.10, $x_1 \equiv_I x_2 \equiv_I x_3 \equiv_I \dots$. Also,

$$\ell(x_{j+1}) = \ell(y_j w_j) \leq \ell(y_j) + \ell(w_j) = \ell(w_j y_j) = \ell(x_j) .$$

Hence, for some N, $\ell(x_N) = \ell(x_{N+1}) = \dots$. Since $y_j \in D_I^{-1}$, we see by the exchange condition that for, $j \geq N$,

$$\begin{aligned} y_{j+1} &= y_j u_j, \ u_j \in W_I \\ \ell(y_{j+1}) &= \ell(y_j) + \ell(u_j) . \end{aligned}$$

In particular, $\ell(y_N) \leq \ell(y_{N+1}) \leq \dots$. Hence, there exists $k \geq N$ such that

$$y_k = y_{k+1} = \dots .$$

Letting $y = y_k$ we obtain, after some scrutiny,

$$y \equiv_I x$$

and

$$\ell(y) \leq \ell(x) .$$

One then checks that this y is unique. See Proposition 1.1 of [90].

We now apply this to our situation. Let $I \subseteq S$. Then for any $J \subseteq S$

$$D_I^{-1} \subseteq (D_I^{-1} \cap D_J) W_J .$$

Thus, for any $y \in D_I^{-1}$, we obtain from standard results that

$$W_I \cap y W_J y^{-1} \text{ is a standard parabolic subgroup of } (W, S).$$

Hence

$$\begin{aligned} W_I \cap y W_I y^{-1} &= W_{I_1} \ , I_1 \subseteq I \\ W_I \cap y W_{I_1} y^{-1} &= W_{I_2} \ , I_2 \subseteq I_1 \\ W_I \cap y W_{I_2} y^{-1} &= W_{I_3} \ , I_3 \subseteq I_2 \\ &\vdots \end{aligned}$$

Then let $K = K_0 \lhd I$, so that K is a union of some connected components of I. Then, as above,

$$\begin{aligned} W_I \cap y W_{K_0} y^{-1} &= W_{K_1} \ , K_1 \lhd I_1 \\ W_I \cap y W_{K_1} y^{-1} &= W_{K_2} \ , K_2 \lhd I_2 \\ W_I \cap y W_{K_2} y^{-1} &= W_{K_3} \ , K_3 \lhd I_3 \\ &\vdots \end{aligned}$$

We arrive at our key definition.

Definition 12.12. *Suppose that $I \subseteq S$ and $K \lhd I$. Define*

$$D_I^*(K) = \{y \in D_I^{-1} \mid y \in D_{K_j} \text{ for all } j \geq 0\}\ .$$

Notice that $D_I^*(\phi) = D_I^{-1}$ and $D_I^*(I) = D_I \cap D_I^{-1}$.

The following corollary follows from Proposition 12.11. See Corollary 1.2 of [90].

Corollary 12.13. *Let $y \in D_I^*(K)$, $z \in W_K$ and suppose that $yz \equiv_I y' \in D_I^{-1}$. Then $\ell(y') \geq \ell(y)$. If further $\ell(y') = \ell(y)$, then $y = y'$.*

We now return to conjugacy classes in M. Let $e \in \Lambda$, the cross-section lattice. Then

$$\begin{aligned} W(e) &= W_I \text{ for } I = \lambda(e) \subseteq S \\ W_*(e) &= \{x \in W \mid xe = ex = e\} = W_K \text{ for some } K \lhd I\ . \end{aligned}$$

Define

$$\begin{aligned} D(e) &= D_I, \\ D^*(e) &= D_I^*(K) \\ D_e &= D_K. \end{aligned}$$

Notice that

$$D^*(e) = D(e) \cap D(e)^{-1} \text{ if } e \in \Lambda \backslash \{0\} \text{ is minimal}$$

and

$$D^*(e) = D(e)^{-1} \text{ if } e \in \Lambda \backslash \{1\} \text{ is maximal}\ .$$

Let $y \in D(e)^{-1}$, $e \in \Lambda$. Define, as before,

$$H = C_G(zez^{-1} \mid z \in \langle y \rangle),$$

and let $M_{e,y} = eHy$. Let

$$X(ey) = \bigcup_{g \in G} gM_{e,\sigma}g^{-1}\ .$$

Theorem 12.14. *Let $e \in \Lambda$.*

a) If $y \in D(e)^{-1}$ then

$$X(ey) = \bigcup_{g \in G} gBeyBg^{-1}\ .$$

b) $GeG = \bigsqcup_{y \in D^*(e)} X(ey)$.

Proof. For a) we refer the reader to Lemma 2.1 and Theorem 2.2 of [90]. For b) let $I = \lambda(e)$. If $x \in W$ then, by Proposition 12.11, $x \equiv_I y$ for some $y \in D(e)^{-1}$ with $\ell(y) \leq \ell(x)$. Thus, by Theorem 12.3 and Theorem 12.9, every element of GeG is conjugate to an element of $M_{e,\sigma}$ for some $y \in D(e)^{-1}$. Furthermore, if $y_1, y_2 \in D(e)^{-1}$ then $X(ey_1) = X(ey_2)$ if and only if for some $x \in W$ $ey_1 \sim ex$ in R and $x \equiv_I y_2$. We write $y_1 \approx y_2$ in this case. If $y_1 \not\approx y_2$ then by Theorem 12.9 $X(ey_1) \cap X(ey_2) = \phi$. Assume that y is chosen so that $\ell(y)$ is minimal in the $\approx$ class of y. It turns out that $y \in D^*(e)$. One then checks that such a y is unique. This proves b).

Definition 12.15. *Let $P = \{ey \mid e \in \Lambda,\ y \in D^*(e)\}$. Since $P \subset R$ we define a transitive relation $\preccurlyeq$ on R: $\preccurlyeq$ is generated by*

a) $r_1 \preccurlyeq r_2$ if $r_1 \leq r_2$ in the adherence order (see Definition 8.32)
b) if $y \in D(e)^{-1}$ and $x \in W$ then $eyx \preccurlyeq xey$.

It follows from Theorem 12.16 a) below that $(P, \preccurlyeq)$ is a partially ordered set. We refer to $(P, \preccurlyeq)$ as the **Putcha poset** of M.

Theorem 12.16. *a) $(P, \preccurlyeq)$ is a poset.*
b) $M = \bigsqcup_{r \in P} X(r)$.
c) If $r_1, r_2 \in P$ then $X(r_1) \subseteq \overline{X(r_2)}$ iff $r_1 \preccurlyeq r_2$.
d) If $r \in P$ then $\overline{X(r)} = \bigsqcup_{s \preccurlyeq r} X(s)$.

Proof. b) follows from Theorem 12.14 b). For c) and d), let $r \in R$ and define

$$Y(r) = \bigcup_{g \in G} gBrBg^{-1} .$$

It turns out that $Y(r) = X(r)$ for any $r \in P$. Now G acts on $Y(r)$ by conjugation, while B stabilizes $\overline{BrB}$ under this action. Thus,

$$\overline{Y(r)} = \bigcup_{g \in G} g\overline{BrB}g^{-1} = \bigcup_{r' \leq r} Y(r')$$

since G/B is a complete variety.

From here one checks that b) and c) hold. For a) it remains to show that if $r_1 \preccurlyeq r_2 \preccurlyeq r_1$ then $r_1 = r_2$. But if $r_1 \preccurlyeq r_2 \preccurlyeq r_1$ then $\overline{X(r_1)} = \overline{X(r_2)}$ by c) above. Hence $X(r_1) \cap X(r_2) \neq \phi$. Thus, by 12.14 $r_1 = r_2$.

Very little is known about the set

$$M_{nil} = \{x \in M \mid x^n = 0 \text{ for some } n\},$$

of "nullforms" of M, except perhaps that $M_n \subseteq M$ is a closed subvariety of M, invariant under the action of conjugation by G. However, by the results above,

$$M_{nil} = \bigsqcup_{r \in P_n} X(r),$$

where $P_{nil} = \{r \in P \mid r^n = 0 \text{ for some } n > 0\}$. Furthermore, the irreducible components of M_{nil} are in one-to-one correspondence with the maximal elements of $(P_{nil}, \preccurlyeq)$.

Example 12.17. Let M be a $\mathcal{J}$-irreducible monoid of type $\phi \subseteq S$. Then by Theorem 7.12,

$$\lambda : \Lambda \backslash \{0\} \longrightarrow 2^S$$

is a bijection. Such monoids are called *canonical monoids* [96]. In any case, we write

$$\Lambda = \{e_X \mid X \subseteq S\} \cup \{0\}\,.$$

The set of maximal elements of Λ^1 of $\Lambda \backslash \{1\}$ are indexed by S. Indeed,

$$\Lambda^1 = \{f_s \mid s \in S\}$$

where $\lambda(f_s) = S \backslash \{s\}$. If $e_X y \in P_n$ then $y = s_1 \cdot \cdots \cdot s_\ell$ where $s_i \in S$ and $s_1 \notin X$. Hence $e_X y \leq f_{s_1} s_1$ and thus $e_X y \preccurlyeq f_{s_1} s_1$. So we see that the set of maximal elements of P_n is

$$\{f_s s \mid s \in s\}\,.$$

Thus the irreducible components of M_{nil} are

$$\overline{X(f_s s)} = \overline{\bigcup_{g \in G} g M_{f_s, s} g^{-1}}\,.$$

Example 12.18. Let $M = M_n(K)$. In this example we calculate the Putcha decomposition and the Putcha poset for M. Now

$$M = \bigsqcup_{r \in P} X(r)$$

as in Theorem 12.16. One checks that for $r \in P$

$$X(r) = \{a \in M \mid \text{ rank } (a^i) = \text{ rank } (r^i) \text{ for all } i > 0\}\,.$$

Furthermore, the following are equivalent for $r, s \in P$:

i) $r \preccurlyeq s$
ii) $X(r) \subseteq \overline{X(s)}$
iii) rank $(r^i) \leq$ rank (s^i) for all $i \geq 0$.

For $r \in P$ we can write

$$r = \left(\begin{array}{c|c} I_r & \\ \hline & N_r \end{array} \right)$$

where I_r is an identity matrix and N_r is nilpotent. Thus,

$$P \cong \bigsqcup_{m \leq n} \Pi_m$$

where Π_m is the set of partitions of m. Here, Π_m corresponds to $\{r \in P \mid m = n - \operatorname{rank}(I_r)\} \subseteq P$.

The order relation on P can be described as follows.

Let

$$\begin{aligned} \alpha &= (\alpha_1 \geq \alpha_2 \geq \ldots) \text{ in } \Pi_m \\ \beta &= (\beta_1 \geq \beta_2 \geq \ldots) \text{ in } \Pi_\ell \,. \end{aligned}$$

Then $\alpha \preccurlyeq \beta$ if

$$\begin{aligned} n - m &\geq n - \ell \\ n - m + \alpha_1 &\geq n - \ell + \beta_1 \\ n - m + \alpha_1 + \alpha_2 &\geq n - \ell + \beta_1 + \beta_2 \\ &\vdots \end{aligned}$$

Each of these inequalities is a direct translation of the corresponding condition from iii) above.

13

The Centralizer of a Semisimple Element

13.1 Introduction

Let G be a simply connected algebraic group and let $s \in G$ be a semisimple element. It is well known that $C_G(s) = \{g \in G \mid gs = sg\}$ is a connected subgroup of G which is uniquely determined up to conjugacy by a certain subset of the extended Dynkin diagram of G.

If M is a reductive monoid with unit group G, the situation is more complicated. Is $C_M(s) = M_s$ always irreducible? If not, can we still obtain some numerical/combinatorial identification of these monoids? What sort of structure does the monoid M_s have?

The purpose of this chapter is to answer the above questions in detail, and to supply some illustrative examples. The three main results are as follows.

Let $B \subseteq G$ be a Borel subgroup with maximal torus $T \subseteq B$. Suppose that $s \in T$ and M is a reductive algebraic monoid with unit group G. Let

$$\begin{aligned} G_s &= \{g \in G \mid gs = sg\} \\ B_s &= \{g \in B \mid gs = sg\} \\ N_s &= \{x \in \overline{N_G(T)} \mid xs = sx\} \end{aligned}$$

and define

$$R_s = N_s/T = \{xT = Tx \mid x \in N_s\} .$$

In this chapter we obtain the following results.

a) $M_s = \bigsqcup_{x \in R_s} B_s x B_s$.
b) M_s is a regular monoid.
c) R_s is a finite inverse monoid.
d) The following are equivalent:
   i) M_s is irreducible
   ii) R_s is unit regular.

13.2 Main Results

Let M be a reductive monoid with unit group G. We assume throughout that G is simply connected. This ensures that, for any semisimple element s of G, $C_G(s) = \{g \in G \mid sg = gs\}$ is connected. So let $B \subseteq G$ be a Borel subgroup with maximal torus $T \subseteq B$. We may assume that $s \in T$.

We now establish our notation and recall the relevant background results. Let

$$N' = \{x \in M \mid xT = Tx\}.$$

Then $N' = N = \overline{N_G(T)} \subseteq M$ (Zariski closure) and $R = \{xT = Tx \mid x \in N\}$ is a finite inverse monoid with unit group $W = N_G(T)/T$, the *Weyl group*. If $x, y \in N$ and $x \equiv y$ in R then $BxB = ByB$. Hence $BxB \subseteq M$ is well-defined for $x \in R$.

Recall from Theorem 8.8 that

$$M = \bigsqcup_{x \in R} BxB.$$

Our purpose here is to find an analogue of this result for

$$M_s = \{x \in M \mid xs = sx\} .$$

So we let

$$\begin{aligned} G_s &= \{x \in M \mid xs = sx\} = C_G(s) \\ B_s &= C_B(s) \\ N_s &= C_N(s) \\ R_s &= \{xT = Tx \in R \mid xT \cap N_s \neq \phi\} . \end{aligned}$$

Notice that if $xT \cap N_s \neq \phi$ then $sxt = xts$ for $t \in T$. It follows easily that $xT \subseteq N_s$. Indeed,

$$R_s = \{xT \in R \mid xT \subseteq N_s\} .$$

Lemma 13.1. *Let $r \in R$. Then $BrB \cong rT \times K^a$ for some $a \geq 0$.*

Proof. Let $V = \{u \in U \mid urB \subseteq rB\}$ where $U \subseteq B$ is the unipotent part of B. Then it follows easily that $V = \{u \in U \mid urB = rB\} = \{u \in U \mid \overline{urB} = \overline{rB}\}$, that $V \subseteq U$ is closed, and that $T \subseteq N_G(V)$. By Proposition 28.1 of [40], $V = \prod_{U_\alpha \subseteq V} U_\alpha$, where $U_\alpha \subseteq U$ is a root subgroup, $\alpha \in \Phi_+$. Let $X = \prod_{U_\alpha \subseteq U} U_\alpha$. Then

$$X \times V \longrightarrow U$$

$$(u, v) \longmapsto xv$$

is an isomorphism Proposition 28.1 of [40] Thus,

$$\begin{aligned} BrB &= UTrB \\ &= UrTB \\ &= UrB \\ &= XVrB \\ &= XrB\,. \end{aligned}$$

Thus $\varphi : X \times rB \longrightarrow BrB$ is surjective. But φ is also injective. Indeed, suppose that $xrb_1 = yrb_2$. Then $rb_1 = x^{-1}yrb_2$ and we obtain $rB = x^{-1}yrB$. Hence $x^{-1}y \in V$, and so $x^{-1}y = v \in V$. Thus $xV = yV$, and so $x = y$ since $x \times V \cong U$. Hence $BrB \cong X \times rB$.

Now let $Z = \{u \in U \mid rTu = rT\}$. As for V, $Z = \prod_{U_\alpha \subseteq Z} U_\alpha$. Then let $Y = \prod_{U_\alpha \not\subseteq Z} U_\alpha$. As above, it follows that

$$rB = rTY \cong rT \times Y\,.$$

We conclude that

$$BrB \cong X \times rB \cong X \times rT \times Y\,.$$

But $X \times Y \cong k^a$ for some $a \geq 0$.

Lemma 13.2. *There is a unique morphism of algebraic varieties $\psi : BrB \to rT$ such that i is the inclusion, α is defined by $\alpha(x) = sxs^{-1}$, β is induced from α and the following diagram commutes.*

$$\begin{array}{ccccc} rT & \xrightarrow{\;i\;} & BrB & \xrightarrow{\;\psi\;} & rT \\ {\scriptstyle\alpha}\big\downarrow & & {\scriptstyle\alpha}\big\downarrow & & {\scriptstyle\beta}\big\downarrow \\ rT & \xrightarrow{\;i\;} & BrB & \xrightarrow{\;\psi\;} & rT \end{array}$$

Furthermore, $\psi \circ i$ is an isomorphism.

Proof. Since $BrB = X \times rT \times Y$, define $\psi(x, rt, y) = rt$. Then ψ is unique since it is the quotient morphism for the action $U \times U \times BrB \to BrB$, $(u, v, x) \mapsto uxv^{-1}$. The diagram commutes as long as β exists. But, for any $t \in T$, $s(UrtU)s^{-1} = UrtU$.

Corollary 13.3. $(BrB)_s = \{b_1rb_2 \in BrB \mid sb_1rb_2s^{-1} = b_1rb_2\}$. *Then*

$$(BrB)_s \neq \phi \Leftrightarrow (rT)_s \neq \phi\,.$$

Proof. Assume that $(BrB)_s \neq \phi$. Then $\psi((BrB)_s) \neq \phi$. But then $(rT)_s \neq \phi$ since $\psi \circ i$ is an $int(s)$-equivariant isomorphism. Conversely, if $(rT)_s \neq \phi$ then $(rT)_s \subseteq (BrB)_s$, and so $(BrB)_s \neq \phi$.

Proposition 13.4.

$$(BrB)_s = \begin{cases} \phi & \textit{if}\, r \notin R_s \\ B_s r B_s & \textit{if}\, r \in R_s. \end{cases}$$

Proof. If $r \notin R_s$ then $(rT)_s = \phi$ by definition, and so $(BrB)_s = \phi$ by Corollary 13.3. Then let $r \in R_s$. We must show that $(BrB)_s = B_s r B_s$. Clearly, $B_s r B_s \subseteq (BrB)_s$. In the proof of Lemma 13.1 we showed that $BrB = X \times rT \times Y$. Then

$$\begin{aligned} (BrB)_s &= (X \times rT \times Y)_s \\ &= X_s \times rT \times Y_s, \text{ since } sXs^{-1} = X \text{ and } sYs^{-1} = Y \\ &\subseteq B_s r B_s. \end{aligned}$$

Theorem 13.5. $M_s = \bigsqcup_{r \in R_s} B_s r B_s$*, a disjoint union.*

Proof.

$$\begin{aligned} M_s &= (\bigsqcup_{r \in R} BrB)_s, \text{ by } 8.8\,, \\ &= \bigsqcup_{r \in R_s} (BrB)_s \\ &= \bigsqcup_{r \in R_s} B_s r B_s, \text{ by } 13.4. \end{aligned}$$

13.3 The Structure of R_s and M_s

In this section we examine in more detail R_s and M_s. But first we recall three definitions from § 2.3. A semigroup S is *regular* if for any $x \in S$ there exists $a \in S$ such that $xax = x$. Also S is *unit regular* if for any $x \in S$ there exists a unit $a \in S$ such that $xax = x$. A semigroup S is *inverse* if for any $x \in S$ there is a unique $x^* \in S$ such that $xx^*x = x$ and $x^*xx^* = x^*$.

Proposition 13.6. R_s *is a finite inverse monoid.*

Proof. $R_s \subseteq R$, which is finite. Hence R_s is finite. Also, it is easily verified that R_s is a semigroup of R. Therefore let $x \in R_0$ and let $r^* \in R$ be the unique inverse (in R) of r. Now $srts^{-1} = rt$ for some $t \in T$. Thus $st^{-1}r^*s^{-1} = (srts^{-1})^* = (rt)^* = t^{-1}r^*$, and so $(r^*T)_s = (Tr^*)_s \neq \phi$, proving that $r^* \in R_s$.

Proposition 13.7. M_s *is a regular, algebraic monoid.*

Proof. By Theorem 13.5 we have $M_s = \bigsqcup_{r \in R_s} B_s r B_s$. Clearly, M_s is a closed submonoid of M. Now let $x = b_1 r b_2 \in M_s$, where $b_1, b_2 \in B_s$ and R_s. Define $a = b_2^{-1} r^* b_1^{-1} \in M_s$ where $r^* \in R_s$ is the unique inverse of r. A simple calculation proves that $xax = x$ and $axa = a$. Thus, M_s is regular.

Theorem 13.8. *The following are equivalent:*

(a) M_s *is irreducible;*
(b) R_s *is unit regular.*

Proof. Recall that $N_s = \{x \in \overline{N_G(T)} \mid xs = sx\}$. A simple calculation verifies that R_s is unit regular iff N_s is unit regular.

Assume that M_s is irreducible. Then $M_s = \overline{C_G(s)}$ (Zariski closure). Now let $r \in R_s$, so that $rs = sr$ and $rT = Tr$. But also we have $r \in M_s$. Hence $r \in \overline{N_{C_G(s)}(T)}$ and so $R_s = \overline{N_{C_G(s)}(T)}$. The latter is unit regular by Theorem 13 of [81] and Theorem 7.3 of [82].

Conversely, assume that N_s is unit regular. Then if $r \in R_s$ there exist $\sigma \in N_{C_G(s)}(T)$ and $e \in I(\overline{T}) = \{f \in \overline{T} \mid f^2 = f\}$ such that $r = e\sigma$. Now let $x \in M_s$. Then $x = b_1 r b_2$, for some $r \in N_s$ and $b_1, b_2 \in B_s$. But $r = e\sigma$ as above, so that $x = b_1 e\sigma b_2 \in B_s\overline{T}N_{C_G(s)}(T)B_s \subseteq \overline{C_G(s)}$. Thus $M_s = \overline{C_G(s)}$.

13.4 Examples

Example 13.9. Let $M = M_n(k)$ and let $s \in M$ be semisimple. Then s is conjugate to a matrix of the form

$$\begin{pmatrix} \lambda_1 I_{n_1} & & 0 \\ & \ddots & \\ 0 & & \lambda_s I_{n_s} \end{pmatrix}$$

where $\lambda_i \neq \lambda_j$ if $i \neq j$, $n_1 \geq \cdots \geq n_s$, and $\sum_{j=1}^{m} n_j = n$. Then

$$M_s \cong \prod_{i=1}^{m} M_{n_i}(k)$$

and

$$R_s \cong \left\{ \begin{pmatrix} A_{n_1} & & 0 \\ & \ddots & \\ 0 & & A_{n_m} \end{pmatrix} \;\middle|\; \begin{array}{l} A_{n_i} \text{ is an } n_i \times n_i \text{ 0-1 matrix with} \\ \text{at most one nonzero entry in each} \\ \text{row or column.} \end{array} \right\}.$$

For this M, any choice of s yields an irreducible M_s.

Example 13.10. Let $\rho : S\ell_2 \times S\ell_2 \to G\ell_6$ be defined by

$$\rho(A, B) = \begin{pmatrix} A \otimes {}^tB^{-1} & 0 \\ 0 & B \end{pmatrix}.$$

Let $G_1 = \rho(S\ell_2 \times S\ell_2)$ and let $G = \{tg \mid t \in ZG\ell_6,\ g \in G_1\}$. Then

$$M = \overline{G} \subseteq M_6(k)$$

is a reductive algebraic monoid with unit group G and maximal torus closure

$$\overline{T} = \left\{ diag(w,x,y,z,r,s) \;\middle|\; \begin{array}{l} wz = xy = rs \\ r^2 = xz \\ s^2 = wy \end{array} \right\} .$$

We can calculate $E(\overline{T})$ to obtain

$$E(\overline{T}) = \{0,1\} \cup \{e_i \mid i = 1, \ldots, 8\}$$

where

$$\begin{array}{ll} e_1 = (1,0,0,0,0,0) & e_5 = (1,1,0,0,0,0) \\ e_2 = (0,1,0,0,0,0 & e_6 = (0,0,1,1,0,0) \\ e_3 = (0,0,1,0,0,0) & e_7 = (1,0,1,0,0,1) \\ e_4 = (0,0,0,1,0,0) & e_8 = (0,1,0,1,1,0). \end{array}$$

It follows from Theorem 10.7 of [82] that the partially ordered set $\{GxG \mid x \in G\}$ is $\{0, J_1, J_2, J_3, G\}$, where

$$J_1 = Ge_1G, \quad J_2 = Ge_5G \quad \text{and} \quad J_3 = Ge_7G .$$

Furthermore, $J_3 > J_1$ and $J_2 > J_1$.

The Weyl group of G is $W = \{w_1, w_2, w_3, w_4\}$ where

$$w_1 = \begin{pmatrix} 1 & & & & & \\ & 1 & & & & \\ & & 1 & & 0 & \\ & & & 1 & & \\ & 0 & & & 1 & \\ & & & & & 1 \end{pmatrix} = \rho\left(\begin{pmatrix} 1 & 0 \\ 0 & 1 \end{pmatrix}, \begin{pmatrix} 1 & 0 \\ 0 & 1 \end{pmatrix}\right)$$

$$w_2 = \left(\begin{array}{cc|cc|cc} 0 & 1 & & & 0 & \\ -1 & 0 & & & & \\ \hline & & 0 & 1 & & \\ & & -1 & 0 & & \\ \hline & 0 & & & 0 & 1 \\ & & & & -1 & 0 \end{array}\right) = \rho\left(\begin{pmatrix} 1 & 0 \\ 0 & 1 \end{pmatrix}, \begin{pmatrix} 0 & 1 \\ -1 & 0 \end{pmatrix}\right)$$

$$w_3 = \left(\begin{array}{cccc|cc} & 0 & & 1 & & \\ & & -1 & & & \\ & -1 & & & 0 & \\ & & 0 & & & \\ 1 & & & & & \\ \hline & 0 & & & 0 & 1 \\ & & & & -1 & 0 \end{array}\right) = \rho\left(\begin{pmatrix} 0 & 1 \\ -1 & 0 \end{pmatrix}, \begin{pmatrix} 0 & 1 \\ -1 & 0 \end{pmatrix}\right)$$

$$w_4 = \left(\begin{array}{cc|cc|cc} 0 & 0 & -1 & 0 & 0 & 0 \\ 0 & 0 & 0 & -1 & 0 & 0 \\ \hline -1 & 0 & 0 & 0 & 0 & 0 \\ 0 & -1 & 0 & 0 & 0 & 0 \\ \hline 0 & 0 & 0 & 0 & 1 & 0 \\ 0 & 0 & 0 & 0 & 0 & 1 \end{array}\right) = \rho\left(\begin{pmatrix} 0 & 1 \\ -1 & 0 \end{pmatrix}, \begin{pmatrix} 1 & 0 \\ 0 & 1 \end{pmatrix}\right).$$

Let $s \in T \subseteq G$ be semisimple. Then

$$s = a\rho(u, v)$$

where $u = diag(\alpha, \alpha^{-1})$ and $v = diag(\beta, \beta^{-1})$. Then

$$s = a\ diag(\alpha\beta^{-1}, \alpha\beta, \alpha^{-1}\beta^{-1}, \alpha^{-1}\beta, \beta, \beta^{-1}).$$

After straightforward but tedious calculations, for example,

$$sw_3s^{-1} = \left(\begin{array}{cccc|cc} 0 & 0 & 0 & \alpha^2\beta^{-2} & 0 & 0 \\ 0 & 0 & -\alpha^2\beta^2 & 0 & 0 & 0 \\ 0 & -\alpha^{-2}\beta^{-2} & 0 & 0 & 0 & 0 \\ \alpha^{-2}\beta^2 & 0 & 0 & 0 & 0 & 0 \\ \hline 0 & 0 & 0 & 0 & 0 & \beta^2 \\ 0 & 0 & 0 & 0 & -\beta^{-2} & 0 \end{array}\right),$$

we arrive at the following possibilities for R_s.

Case 1: $\beta = \pm\alpha$, $\alpha = \pm 1$.

$$\text{Then } R_s = R\ .$$

Case 2: $\beta = \pm\alpha$, $\alpha = \pm i$.

$$\text{Then } R_3 = E(\overline{T}) \cup \{w_3e_j \mid j = 1, 2, \ldots, 6\}\ .$$

Case 3: $\beta = \pm\alpha$ and $\alpha \neq \pm 1$, $\pm i$.

$$\text{Then } R_3 = E(\overline{T}) \cup \{w_3e_1, w_3e_4\}\ .$$

Case 3′: $\beta = \pm\alpha^{-1}$, $\alpha \neq \pm 1$, $\pm i$ gives a set R_s conjugate to the set R_s considered in Case 3.

Case 4: $\beta \neq \pm\alpha$, $\alpha = \pm 1$.

$$\text{Then } R_s = E(\overline{T}) \cup w_4(E(\overline{T})\ .$$

Case 4′: $\beta \neq \pm\alpha^{-1}$, $\beta = \pm 1$ gives a set R_s conjugate to the set R_s considered in Case 4.

Case 5: $\beta \neq \pm\alpha$, $\beta \neq \pm\alpha^{-1}$, $\alpha \neq \pm 1$, $\beta \neq \pm 1$.

$$\text{Then } R_s = E(\overline{T})\,.$$

In Case 1, $M_s = M$.
In Case 4 or Case 4′, M_s is irreducible with unit group $k^* S\ell_2$.
In Case 5, $M_s = \overline{T}$.
In Case 2, Case 3 and Case 3′, M_s is reducible.

Remark 13.11. The monoid M_s is not necessarily of the type discussed in [85, 86] or § 10.4, unless of course it is irreducible. This leads to a number of basic questions about M_s.

a) Which spherical varieties (for $C_G(s) \times C_G(s)$) can occur as an irreducible component of M_s? Are there any "new" ones?
b) Does the inverse monoid R_s satisfy some analogue of Tits' axiom "$sBx \subseteq BxB \cup BsxB$ if $s \in S$, $x \in R$"?
c) Is there an analogue of the type map $\lambda : \Lambda \to 2^S$ for M_s?

14

Combinatorics Related to Algebraic Monoids

In this section, we discuss some of the more striking combinatorial problems that arise naturally in the study of reductive, algebraic monoids.

14.1 The Adherence Order on WeW

From Corollary 8.35 we have a combinatorial description of the adherence ordering on R:

$$x \leq y \text{ if } BxB \subseteq \overline{ByB}.$$

In this section we describe a refinement of those results when we restrict our attention to the smaller poset $(WeW, \leq)$ for $e \in \Lambda$. The results of this section are due to Putcha [92, 93, 94].

The monoid R has a presentation given by:

$$\begin{aligned} xe &= ex, \qquad x \in W(e) \\ xe &= e = ex, \; x \in W_*(e). \end{aligned}$$

These relations are of course in addition to those for W and $E(\overline{T})$.

If $\sigma \in R$, then σ has a unique expression:

$$\sigma = xwey, e \in \Lambda, x \in D(e), w \in W^*(e), y \in D(e)^{-1}.$$

This is the *normal form* (Definition 8.34) of σ. The *length* $l(\sigma)$ can be defined (see [73], [105], [127]) as

$$l(\sigma) = l(x) + l(w) + l(e) - l(y)$$

where $l(e)$ is the length of the longest element in $D(e)$. This notion of length agrees with the earlier Definition 8.17. If v_0 and w_0 are respectively the longest elements in $W(e)$ and W, then w_0e and ev_0w_0 are respectively the maximum and minimum elements of WeW. Moreover the length function on WeW agrees with the rank function on the graded poset WeW. Also by [73], the length function is subadditive:

$$l(\sigma\theta) \leq l(\sigma) + l(\theta) \quad \text{for all} \quad \sigma, \theta \in \mathrm{R}.$$

Let $\sigma = xwey, \sigma' = x'w'e'y' \in R$ be in standard form. Then by [73],

$$\sigma \leq \sigma' \Leftrightarrow e \leq e', xw \leq x'w'u, u^{-1}y' \leq y \quad \text{for some} \quad u \in W(e')W_*(e).$$

This description of the order on R is somewhat unwieldly. A much more useful description has been obtained by the Putcha in [92, 93], which we now describe.

We begin with a description of the order on WeW, where $e \in \Lambda$. Let $I = \lambda(e)$ and $K = \lambda_*(e)$. For $w \in W_I$, let $\overline{w} = v_0 w v_0$, where v_0 is the longest element of W_I. Let

$$\mathcal{W}_{I,K} = D_I \times W_{I \setminus K} \times D_I^{-1}.$$

Let $\sigma = (x, w, y), \sigma' = (x', w', y',) \in \mathcal{W}_{I,K}$. Define $\sigma \leq \sigma'$ if $w = w_1 w_2 w_3$ with $l(w) = l(w_1) + l(w_2) + l(w_3)$ such that

$$xw_1 \leq x', w_2 \leq w', \overline{w}_3 y \leq y'.$$

Let $\mathcal{W}^*_{I,K} = \mathcal{W}_{I,K}$ as sets but with the above order changed to

$$xw_1 \leq x', w_2 \leq w', w_3 y \leq y'.$$

Note the subtle difference between these two orderings. The following result is proved in [92].

Theorem 14.1. *WeW is isomorphic to $\mathcal{W}_{I,K}$ and the dual of WeW is isomorphic to $\mathcal{W}^*_{I,K}$.*

We notice that, in many cases, the poset WeW is isomorphic to its dual. This is however not true in general. For instance this is not true of the poset of 4×4 rank 3 partial permutation matrices. In general WeW is isomorphic to its dual if no component of $W^*(e)$ is of type $A_l (l > 1), D_l (l \text{ odd})$ or E_6.

The problem next is to extend the description of the order on the $W \times W$-orbits to all of R. Let $e, f \in \Lambda, e \leq f$. Let z_e denote the longest element in $W_*(e)$. Let $\sigma = xwey \in WeW$ in standard form. Let

$$z_e y = u y_1, u \in W(f), y_1 \in D(f)^{-1}.$$

Define the **projection**,

$$\rho_{e,f}(\sigma) = x' f y_1$$

where

$$x' = \min\{xwu' | u' \leq u\}.$$

Thus we have projection maps $\rho_{e,f} : WeW \to WfW$, whenever $e \leq f$.

Example 14.2. Let $M = M_3(k)$,

$$e = \begin{bmatrix} 1 & 0 & 0 \\ 0 & 0 & 0 \\ 0 & 0 & 0 \end{bmatrix}, \quad f = \begin{bmatrix} 1 & 0 & 0 \\ 0 & 1 & 0 \\ 0 & 0 & 0 \end{bmatrix}.$$

Then $\rho_{e,f} : WeW \to WfW$ is as in the Figure 14.1.

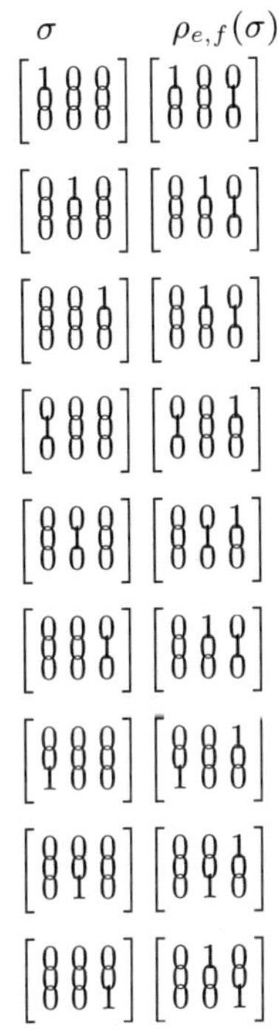

Fig. 14.1. The projection in $M_n(K)$ from rank one to rank two.

The following result is proved in [92].

Theorem 14.3. *(i)* $\rho_{e,f} : WeW \to WfW$ *is order preserving.*

(ii) If $\sigma \in WeW, \theta \in WfW$, *then* $\sigma \leq \theta \Leftrightarrow \rho_{e,f}(\sigma) \leq \theta$.

(iii) $\rho_{e,f}$ *is onto* $\Leftrightarrow \lambda_*(e) \subseteq \lambda_*(f)$.

(iv) $\rho_{e,f}$ *is one to one* $\Leftrightarrow \lambda(f) \subseteq \lambda(e)$.

Example 14.4. Let $\phi : M_n(K) \to M_N(K)$ be defined as

$$\phi(A) = A \otimes \wedge(A) \otimes \wedge^2 \cdots \otimes \wedge^n(A),$$

where

$$N = \Pi_{r=1}^{n} \binom{n}{r}.$$

Let M denote the Zariski closure of $\phi(M_n(K))$ in $M_N(K)$. In this case, W is the symmetric group on n letters with $S = \{(12), (23), \ldots, (n-1n)\}$. Then

$$\Lambda = \{e_I \mid I \subseteq S\} \cup \{0\}$$

with

$$e_K \leq e_I \text{ if and only if } K \subseteq I$$

and

$$\lambda(e_I) = \lambda^*(e_I) = I, \; \lambda_*(e_I) = \phi$$

for all $I \subseteq S$. Then by Theorem 14.3, ρ_{e_K, e_I} is onto for $K \subseteq I$.

Example 14.5. Let $\phi : M_n(K) \to M_N(K)$ be defined as

$$\phi(A) = A \oplus \wedge(A) \oplus \wedge^2 \cdots \oplus \wedge^n(A)$$

where $N = 2^n - 1$. Let M denote the Zariski closure of $\phi(M_n(K)$ in $M_N(K)$. Again W is the symmetric group on n letters with $S = \{(12), (23), \ldots, (n-1n)\}$. Then

$$\Lambda = \{e_I \mid I \subseteq S\} \cup \{1\}$$

with

$$e_K \leq e_I \text{ if and only if } K \subseteq I$$

and

$$\lambda(e_I) = \lambda_*(e_I) = I, \; \lambda^*(e_I) = \phi$$

for all $I \subseteq S$. Again by Theorem 14.3, ρ_{e_K, e_I} is one to one for $K \subseteq I$.

If M is a canonical monoid as in Example 10.17, then Theorem 14.3 can be used to show that $R^* = R \backslash \{0\}$ is an **Eulerian poset**. This means that the Möbius function μ on R^* is given by

$$\mu(x, y) = (-1)^{rkx + rky}, for x \leq y$$

Here rk is the rank function on the graded poset R^*. The problem of determining the Möbius function on R in general, remains open.

14.2 Shellability and Stanley-Reisner Rings

Let P be a finite graded partially ordered set with a maximum element 1 and a minimum element 0. This means that all maximal chains in P have the same length. Let Δ denote the order complex of P. Thus Δ is the simplicial complex whose faces are chains in P. Then Δ (or P) is said to be **shellable** if the maximal faces can be ordered $F_1, \cdots, F_s$ such that, for $i < j$, there is a maximal face $F_k, k < j$, such that $F_i \cap F_j \subseteq F_k \cap F_j = F_j \backslash \{v\}$ for some $v \in P$. This is a topological condition implying that Δ then has the homotopy type of a wedge of spheres. Let $\mathcal{A}$ denote the commutative algebra (over a field) generated by $x_\alpha (\alpha \in P)$. For $A = \{a_1, \cdots, a_m\} \subseteq P$, let $x_A = x_{a_1} \cdots x_{a_m}$. Let $\mathcal{I}$ denote the ideal generated by x_A, where A is not a face. Then $\mathcal{F} = \mathcal{A}/\mathcal{I}$ is called the **Stanley-Reisner ring** of Δ (or P). Note that $\mathcal{F}$ is spanned by the faces of Δ. It is well known that, if Δ is shellable, then $\mathcal{F}$ is a Cohen-MacAulay ring, cf. [132].

A very useful method for checking the shellability of P has been developed by Björner and Wachs [6]. For $a, b \in P$, write $a \to b$ if a covers b (i.e. $a > b$ and there is no c such that $a > c > b$). The concept of **lexicographic shellability**, introduced in [6], can be briefly described as follows. The edges of P are labeled recursively starting from the top, whereby for $a \to b$ the label depends on the choice of a maximal chain from 1 to a. Fix $a > b$ and a maximal chain from 1 to a. The labeling must be such that there exists a unique maximal chain from a to b with increasing labels and so that this chain is lexicographically less than any other maximal chain from a to b. It is shown in [6] that lexicographic shellability implies shellability.

Proctor [80] and Björner and Wachs [6] have shown that the Weyl group W, with respect to the Bruhat-Chevalley order, is lexicographically shellable. It is also shown in [6] that D_I is lexicographically shellable for any $I \subseteq S$.

The following result is due to Putcha [93, 94].

Theorem 14.6. *Each $W \times W$-orbit WeW is lexicographically shellable, and hence its Stanley-Reisner ring is Cohen-Macaulay. A maximal $W \times W$-orbit is also Eulerian and hence its Stanley-Reisner ring is Gorenstein.*

The problem of whether the Stanley-Reisner ring of R is always Cohen-Macaulay, remains open.

14.3 Distribution of Products in Finite Monoids

In this section we consider the problem of distribution of products in a finite monoid S. In our discussion we naturally arrive at formulas that are familiar in enumerative combinatorics. This problem becomes particularly interesting if S is a finite monoid of Lie type. We obtain explicit formulas for the monoid $M_n(\mathbb{F}_q)$.

Let S be a finite monoid and let $a \in S$. Consider,

$$H_n(a) = \{(x_1, \ldots, x_n) \in S^n \mid x_1 \cdot \ldots \cdot x_n = a\}$$

for $n > 0$, and define

$$h_n(a) = |H_n(a)| \ .$$

What can we say about the sequence $\{h_n(a) \mid n > 0\}$? What special properties does the power series

$$h(a) = \sum_{n \geq 1} h_n(a) t^{n-1}$$

possess? If $a = bx$, how are $h(a)$ and $h(b)$ related?

In this section we treat the above questions (and some others) systematically from a combinatorial viewpoint.

14.3.1 Properties of $h(a)$

Let S be a finite monoid and let $a \in S$. As above, we let

$$h_n(a) = \left|\{(x_1, \ldots, x_n) \in S^n \mid x_1 \cdot \ldots \cdot x_n = a\}\right| \ .$$

For $a, b \in S$ we write $b \geq a$ if $a = bx$ for some $x \in S$. We define

$$R(b/a) = \{x \in S \mid a = bx\}$$

and

$$r(b/a) = |R(b/a)| \ .$$

Proposition 14.7. *(i) $h_1(a) = 1$ for any $a \in S$.*
(ii) $h_{n+1}(a) = \sum_{b \geq a} r(b/a) h_n(b)$ for any $n > 1$.

Proof. (i) is obvious.

For (ii) consider

$$\begin{aligned} &\{(x_1, \ldots, x_n, x_{n+1} \in S^{n+1} \mid x_1 \cdot \ldots \cdot x_{n+1} = a\} \\ &= \bigsqcup_{b \in S} \left\{ (x_1, \ldots, x_n, x_{n+1}) \in s^{n+1} \middle| \begin{matrix} x_1 \cdot \ldots \cdot x_{n+1} = a \\ x_1 \cdot \ldots \cdot x_n = b \end{matrix} \right\} \\ &\cong \bigsqcup_{b \geq a} \left(\{(x_1, \ldots, x_n) \in S^n \mid x_1 \cdot \ldots \cdot x_n = b\} \times R(b/a)\right) \ . \end{aligned}$$

Hence $H_{n+1}(a) = \bigsqcup_{b \geq a} H_n(b) \times R(b/a)$, and thus

$$h_{n+1}(a) = \left|H_{n+1}(a)\right| = \sum_{b \geq a} \left|H_n(b)\right| \left|r(b/a)\right| = \sum_{b \geq a} h_n(b) r(b/a) \ .$$

Remark 14.8. One can deduce from this that

$$h_{n+1}(a) = \sum_{a=a_0 \leq \cdots \leq a_n} r(a_n/a_{n-1}) \cdot \ldots \cdot r(a_1/a_0) \quad \text{for any } a \in S\,.$$

For $a, b \in S$ we write $a \sim b$ if $aS = bS$, i.e. if $a \leq b$ and $b \leq a$. Then we obtain from Proposition 14.7 that

$$h(a) - 1 = \sum_{n \geq 1} h_{n+1}(a)t^n = \sum_{n\geq 1} t \sum_{b \geq a} r(b/a)h_n(b)t^{n-1} = t\sum_{b\geq a} r(b/a)h(b)\,.$$

Thus

$$h(a) - t\sum_{b\sim a} r(b/a)h(b) = 1 + t\sum_{b>a} r(b/a)h(b)\,. \qquad (*)$$

Define

$$C_a = 1 + t\sum_{b>a} r(b/a)h(b),$$

and let

$$R_a = \{b \in S \mid a \sim b\}\,.$$

If $R_a = \{a_1, \ldots, a_s\}$ write C_i for C_{a_i}. Then from $(*)$ above

$$\begin{array}{lllll}
(1 - r(a_1/a_1)t)h(a_1) & - \quad r(a_2/a_1)th(a_2) & - \cdots - & r(a_s/a_1)th(a_s) & = C_1 \\
-r(a_1/a_2)th(a_1) & + (1 - r(a_1/a_2)t)h(a_2) & - \cdots - & r(a_s/a_2)th(a_s) & = C_2 \\
\vdots & & & & \vdots \\
-r(a_1/a_s)th(a_1) & - \quad \cdots - \cdots & + & (1 - r(a_s/a_s)t)h(a_s) & = C_s.
\end{array}$$

Hence

$$(I_s - tR)\begin{pmatrix} h(a_1) \\ \vdots \\ h(a_s) \end{pmatrix} = \begin{pmatrix} C_1 \\ \vdots \\ C_s \end{pmatrix} \qquad (**)$$

where

$$R = \begin{pmatrix} r(a_1/a_1) & r(a_2/a_1) & \ldots \\ r(a_1/a_2) & r(a_2/a_2) & \\ & \vdots & \end{pmatrix}.$$

But $(I_s - tR)^{-1} = \frac{1}{det(I_s - tR)}\, Adj(I_s - tR)$ and so we obtain the following result.

Theorem 14.9. *For each $a_i \in R_a = \{c | c \sim a\} = \{a_1, \ldots, a_s\}$.*

$$h(a_i) = \frac{1}{det(I_s - tR)}\left[P_{a_i} + t\sum_{b>a_i} P(b/a_i)h(b)\right]$$

where

$$R = \begin{pmatrix} r(a_1/a_1) & r(a_2/a_2) \dots \\ r(a_1/a_2) & r(a_2/a_2) \\ & \vdots \end{pmatrix},$$

$$\begin{pmatrix} P_{a_1} \\ \vdots \\ P_{a_s} \end{pmatrix} = Adj(I_s - tR) \begin{pmatrix} 1 \\ \vdots \\ 1 \end{pmatrix}$$

and

$$\begin{pmatrix} P(b/a_1) \\ \vdots \\ P(b/a_s) \end{pmatrix} = Adj(I_s - tR) \begin{pmatrix} r(b/a_1) \\ \vdots \\ r(b/a_s) \end{pmatrix}.$$

Proof. From (∗∗)

$$(I_s - tR) \begin{pmatrix} h(a_1) \\ \vdots \\ h(a_s) \end{pmatrix} = \begin{pmatrix} 1 + t \sum_{b>a_1} r(b/a_1)h(b) \\ \vdots \\ 1 + t \sum_{b>a_s} r(b/a_s)h(b) \end{pmatrix}$$

and so

$$\begin{pmatrix} h(a_1) \\ \vdots \\ h(a_s) \end{pmatrix} = (I_s - tR)^{-1} \begin{pmatrix} 1 + t \sum_{b>a_1} r(b/a_1)h(b) \\ \vdots \\ 1 + t \sum_{b>a_s} r(b/a_s)h(b) \end{pmatrix}.$$

The result follows from straightforward calculation.

Suppose now that S has the property "$a \sim b \Rightarrow a = b$". We then refer to S as an $\mathcal{R}$-*trivial monoid.* In this case, $s = 1$ for each a, and $I_s - tR = \frac{1}{1-r(a/a)t}$. Then we obtain the following special case.

Corollary 14.10. *Suppose that S is an $\mathcal{R}$-trivial monoid. Then for any $a \in S$*

$$h(a) = \frac{1}{1 - r(a/a)t} \left(1 + t \sum_{b>a} r(b/a)h(b) \right).$$

Example 14.11. Let $S = J \cup \{0\}$ be a completely 0-simple finite semigroup (not usually a monoid, but h_n is still defined). Then by the Rees representation theorem

$$J = I \times G \times \Lambda .$$

Here G is a finite group, and $P : \Lambda \times I \longrightarrow G \cup \{0\}$ is the sandwich matrix. Assume that

$$|I| = k$$
$$|\Lambda| = \ell$$
$$|G| = m.$$

and suppose that

$$|\{(\lambda, i) \in \Lambda \times I \mid p(\lambda, i) = 0\}| = \alpha .$$

Then for any $a \in J$

$$h_n(a) = [g(k\ell - \alpha)]^{n-1} .$$

Hence

$$h(a) = \frac{1}{1 - g(k\ell - \alpha)t} .$$

Example 14.12. Let $F = F(X)$ be the free monoid on the set X. If $a \in F$ then we can write

$$a = x_1 \cdot \ldots \cdot x_n$$

uniquely, with $\{x_i\} \subseteq X$. Hence

$$\{b \in F \mid b \geq a\} = \{1, x_1, x_1x_2, \ldots, x_i \cdot \ldots \cdot x_n\}$$

and

$$r(b/a) = 1 \quad \text{for any} \quad b \geq a .$$

So, inductively,

$$\begin{aligned} h(a) &= \frac{1}{1-t}\left(1 + t\sum_{i=0}^{n-1} h(x_i \cdot \ldots \cdot x_i)\right) \\ &= \frac{1}{1-t}\left(1 + \frac{t}{1-t} + \cdots + \frac{t}{(1-t)^n}\right) \\ &= \frac{1}{(1-t)^{n+1}} . \end{aligned}$$

Remark 14.13. (i) Theorem 14.9 yields an upper bound to the poles of $h(a)$ in terms of the matrices

$$R(b) = \begin{pmatrix} r(b_1/b_1 & r(b_2/b_1) & \ldots \\ r(b_1/b_2) & r(b_2/b_2) & \\ \vdots & & \end{pmatrix}$$

where $\{b_1, \ldots, b_s\} = \{x \in S \mid x \sim b\}$ and $b \geq a$. I am not aware of any example where the poles of $h(a)$ are not reciprocals of integers.

(ii) One could apply the ideas of this section almost verbatim to finite categories.

Definition 14.14. *A finite monoid S is called* $\mathcal{R}$-homogeneous *if*

(i) $r(a_1/a_2) = r(a_3/a_4)$ *if* $a_1 \sim a_2 \sim a_3 \sim a_4$
(ii) $r(b/a_1) = r(b/a_2)$ *if* $b > a_1 \sim a_2$.

Theorem 14.15. *Let S be a finite $\mathcal{R}$-homogeneous monoid. Then for $a \in S$*

$$h(a) = \frac{1}{1 - s_a r_a t}\left(1 + t\sum_{b>a} r(b/a)h(b)\right)$$

where $r = r_a = r(a/a)$ and $s_a = |\{x \in S \mid x \sim a\}|$.

Proof. Let $\{x \in S \mid x \sim a\}$ $\{a_1, \ldots, a_s\}$ and define

$$R = \begin{pmatrix} r(a_1/a_1) & r(a_2/a_1) \ldots \\ r(a_1/a_2) & r(a_1/a_1) \\ & \vdots \end{pmatrix}.$$

Then $rank\ (R) = 1$ and $R^2 = rsR$ (since $r = r(a_i/a_j)$ for all i, j). Then R is conjugate to

$$A = \begin{pmatrix} rs & & \\ & 0 & 0 \\ 0 & & \ddots \\ & & 0 \end{pmatrix}.$$

Thus, $det(I - tR) = det(I - tA) = 1 - rst$. Also

$$\begin{aligned} (I - tR)^{-1} &= I + tR + t^2R^2 + \ldots \\ &= I + tR + t^2 rsR + t^3(rs)^2 R + \ldots \\ &= I + tR(I + rst + (rst)^2 + (rst)^3 + \ldots) \\ &= I + \frac{t}{1 - rst} R\,. \end{aligned}$$

Thus $adj(I - tR) = (1 - rst)I + tR$. Hence, by calculation,

$$Adj(I - tR)\begin{pmatrix} 1 \\ \vdots \\ 1 \end{pmatrix} = \begin{pmatrix} 1 \\ \vdots \\ 1 \end{pmatrix}$$

and

$$Adj(I - tR)\begin{pmatrix} r(b/a_1) \\ \vdots \\ r(b/a_s) \end{pmatrix} = r(b/a)\begin{pmatrix} 1 \\ \vdots \\ 1 \end{pmatrix}$$

since $r(b/a_i) = r(b/a_j)$. Thus, by Theorem 14.9,

$$h(a) = \frac{1}{1 - rst}\left(1 + t\sum_{b>a} r(b/a)h(b)\right).$$

Example 14.16. Let S be a finite monoid such that if $a \sim a'$ then $a' = ag$ for some invertible element $g \in S$ (say $g \in G(S)$). Hence if $a_1 \sim a_2 \sim a_3 \sim a_4$ then

(i) $R(a_4/a_3) = k^{-1}R(a_2/a_1)h$ if $a_3 = a_1h$ and $a_4 = a_2k$
(ii) $R(b/a') = R(b/a)g$ if $b > a$ and $a' = ag$.

It follows that S is $\mathcal{R}$-homogeneous.

The finite monoids of Lie type are a class of finite monoids which satisfy the $\mathcal{R}$-homogeneous property (because of the above property). It would be interesting to obtain detailed information about $h(a)$ in for such monoids. See Example 14.3.2 below for a particular case.

Example 14.17. A finite monoid S that is not $\mathcal{R}$-homogeneous.

Let

$$i = \begin{pmatrix} 1&0&0\\ 0&1&0\\ 0&0&1 \end{pmatrix}, \quad c = \begin{pmatrix} 1&0&0\\ 0&0&0\\ 0&0&0 \end{pmatrix}, \quad f = \begin{pmatrix} 1&1&0\\ 0&0&0\\ 0&0&0 \end{pmatrix}, \quad x = \begin{pmatrix} 1&0&0\\ 0&0&0\\ 0&1&0 \end{pmatrix}.$$

Then $S = \{I, e, f, x\}$ is a finite monoid with $e \sim f < x$. But

$$R(x/e) = \{e, x\}$$

and

$$R(x/f) = \{f\}.$$

Thus, $r(x/e) = 2 \neq r(x/f) = 1$. So by (ii) of Definition 14.14, S is not $\mathcal{R}$-homogeneous.

14.3.2 Example

In this example we consider the example $S = M_n(\mathbb{F}_q)$ in detail, where $\mathbb{F}_q$ is the field with q elements.

Let $a \in S$. Then, for some $g, h \in G(S)$,

$$gah = e_i = \begin{pmatrix} 1 & & & & & \\ & \ddots & & & 0 & \\ & & 1 & & & \\ & & & 0 & & \\ & 0 & & & \ddots & \\ & & & & & 0 \end{pmatrix}$$

for some $0 \le i \le a$ (where $i = \text{rank}\,(e_i)$). It follows easily that

$$r(a/a) = r(e_i/e_i) = \left|\{x \in S \mid e_i = e_iX\}\right|.$$

An elementary calculation verifies that

$$r_i = r(e_i/e_i) = q^{n(n-i)} .$$

Next we need to find

$$s_i = \left|\{b \in S \mid b \sim e_i\}\right| = \left|e_i G\ell_n(\mathbb{F}_q)\right| .$$

Another simple calculation yields

$$s_i = (q^n - 1)(q^n - q) \cdot \ldots \cdot (q^n - q^{n-1}) .$$

Hence

$$s_i r_i = q^{n(n-i)}(q^n - 1)(q^n - q) \cdot \ldots \cdot (q^n - q^{n-1}) .$$

Notice that $s_i r_i \neq s_j r_j$ if $i \neq j$. Thus by 3.4 above, α is semisimple, and so

$$min(\alpha) = X \prod_{i=0}^{n} (X - q^{n(n-i)}(q^n - 1) \cdot \ldots \cdot (q^n - q^{n-1}))$$

and

$$D = \prod_{i=0}^{n} (1 - q^{n(n-i)}(q^n - 1) \cdot \ldots \cdot (q^n - q^{n-1})X).$$

Furthermore, by Theorem 14.15,

$$h(e_i) = \frac{1}{1 - r_i s_i t} \left(1 + t \sum_{b > e_i} r(b/e_i) h(b)\right) .$$

We use this to find an explicit formula of the form

$$h(e_i) = \frac{1}{1 - r_i s_i t} \left(1 + t \sum_{j > i} A_i \; h(e_i)\right) .$$

We first calculate

$$r(b/e_i) = \left|\{x \in S \mid bx = e_i\}\right| .$$

Suppose that $bx = e_i$. Then

$$\{x \mid bx = e_i\} = x + \{y \mid by = 0\} .$$

Thus $r(b/e_i) = q^d$ where $d = dim_{\mathbb{F}_q}(\{y|by = 0\})$. An elementary calculation shows that $d = n(n-j)$, where $j = \text{rank}(b)$. To find the sought after formulas, it remains to find

$$\left|\left\{ b \in S \,\middle|\, \begin{array}{l} bx \;\; = \;\; e_i \;\; \text{for some} \;\; x \\ \text{rank}(b) = j \end{array} \right\}\right| \qquad \text{for each } j > i .$$

Thus we define

$$X_j = \left\{ b \in S \,\middle|\, \begin{array}{l} bx = e_i \text{ for some } x \text{ and} \\ \text{rank } (b) = j \end{array} \right\} \quad \text{for each } j > i \,.$$

Define

$$P_i = \{g \in G \,|\, ge_i = e_i g\, e_i\}$$

where

$$G = G\ell_n(\mathbb{F}_q)\,.$$

Notice also that

$$P_i X_j G = X_j \quad \text{for all } j > i\,.$$

After a little calculation we conclude the following.

Let

$$\mathcal{R}_n = \left\{ A \in M_n(\mathbb{F}_q) \,\middle|\, \begin{array}{l} A \text{ is a 01 matrix with at most one} \\ \text{nonzero entry in each row or col-} \\ \text{umn} \end{array} \right\}.$$

and let

$$\mathcal{X}_j = \{r \in \mathcal{R}_n \mid e_i \in r\mathcal{R}_n,\ \text{rank } (r) = j\}\,.$$

Then

$$X_j = P_i \mathcal{X}_j\, G\,.$$

Now

$$\mathcal{X} =: \bigcup_{j>i} \mathcal{X}_j = \bigcup_{e \in E(\mathcal{X})} eW$$

and after a little more calculation we obtain

$$\mathcal{X} = C_W(e_i)\Lambda(\mathcal{X})W$$

where $W \subseteq \mathcal{R}_n$ is the unit group and

$$\Lambda(\mathcal{X}) = \{e_{i+1}, e_{i+2}, \ldots, e_n = 1\}\,.$$

Thus,

$$X_j = P_i e_j G$$

and

$$X = \bigcup_{j=i+1}^{n} X_j\,.$$

But

$$|X_j| = (q^j - 1) \cdot \ldots \cdot (q^j - q^{j-1}) \begin{bmatrix} u \\ j \end{bmatrix}_q \begin{bmatrix} n-i \\ j-i \end{bmatrix}_q$$

where

$$\begin{bmatrix} n \\ j \end{bmatrix}_q = \frac{(q^n - 1) \cdot \ldots \cdot (q-1)}{[(q^j - 1) \cdot \ldots \cdot (q-1)][(q^{n-k} - 1) \cdot \ldots \cdot (q-1)]}\,.$$

Hence

$$
\begin{aligned}
h(e_i) &= \frac{1}{1 - r_i s_i t}\left(1 + t \sum_{b > e_i} r(b/e_i) h(b)\right) \\
&= \frac{1}{1 - r_i s_i t}\left(1 + t \sum_{j=i+1}^{n} \sum_{\substack{b > e_i \\ \text{rank } (b) = j}} r(b/e_i) h(b)\right) \\
&= \frac{1}{1 - r_i s_i t}\left(1 + t \sum_{j=i+1}^{n} q^{n(n-j)} \left|\left\{ b \,\middle|\, \begin{matrix} bx = e_i \text{ for some } x \text{ rank} \\ (b) = j \end{matrix} \right\}\right| h(b)\right) \\
&= \frac{\left(1 + t \displaystyle\sum_{j=i+1}^{n} q^{n(n-j)} (q^j - 1) \cdot \ldots \cdot (q^j - q^{j-1}) \begin{bmatrix} n \\ j \end{bmatrix}_q \begin{bmatrix} n-i \\ j-i \end{bmatrix}_q h(e_j)\right)}{1 - q^{n(n-i)} (q^n - 1) \cdot \ldots \cdot (q - 1) t} .
\end{aligned}
$$

14.4 Exercises

1. Let $M = M_n(K)$, $R_n = \overline{N_G(T)} \subseteq M$ and $R_n = R/T$. Let $r_n = |R_n|$.
   a) Prove that $r_0 = 1$, $r_1 = 2$ and $r_n = 2nr_{n-1} - (n-1)^2 r_{n-2}$ for $n \geq 2$.
   b) Let $r(x) = \sum_{n=0}^{\infty} \frac{r_n}{n!} x^n$. Prove that $\frac{r'(x)}{r(x)} = \frac{2-x}{(1-x)^2}$.
   c) Show that $r(x) = \frac{e^{x/(1-x)}}{1-x}$.
   See [8].

15

Survey of Related Developments

In this chapter we describe several results that are directly related to the theory of algebraic monoids. In each case we find some other areas of algebra coming into play. Although many of these technical issues are beyond the scope of this survey, we hope to provide the reader with an overview of some of the more striking related developments.

15.1 Complex Representation of Finite Reductive Monoids

As is well known, a finite group G has the property that every finite dimensional complex representation of G is completely reducible. The situation for finite monoids is entirely different and not nearly so well understood. Indeed, there is no effective characterization of finite monoids whose complex representations are all completely reducible. Until recently, it was not known that $\mathbb{C}[M_n(\mathbb{F}_q)]$ is a semisimple algebra.

The following result was made possible by the theory of reductive monoids. It is due to Okninski and Putcha [67].

Theorem 15.1. *Let M be a finite monoid of Lie type. Then the monoid algebra $\mathbb{C}[M]$ is semisimple.*

Recall that a *monoid of Lie type* is a regular monoid M generated by idempotents and units, whose unit group is a group of Lie type. These monoids are defined to satisfy a number of axioms relating to the BN-pair structure of G to the idempotents of M. See Theorem 10.4. In particular, any finite reductive monoid is a finite monoid of Lie type. The model example here is $M_n(\mathbb{F}_q)$.

Corollary 15.2. *Let M be as in the theorem. Then $\mathbb{C}[M_n(\mathbb{F}_q)]$ is a semisimple algebra.*

This result was also obtained independently by Kovacs [49].

As a consequence of their approach to Theorem 15.1, the authors obtain the following important corollary.

Corollary 15.3. *Any complex irreducible representation of M is induced from some irreducible representation $\rho : P \to G\ell_n(\mathbb{C})$ of a parabolic subgroup $P \subseteq G$ with $R_u(P) \subseteq Ker(\rho)$.*

The proof combines two opposing strategies.

a) Harish-Chandra's theory of cuspidal representations,
b) reduction to certain monoids of the form $M = G \cup GeG \cup \{0\}$.

In a subsequent paper [68] the authors turn the tables. They *assume* that M is a finite monoid with unit group G and zero element $0 \in M$, and semisimple monoid algebra $\mathbb{C}[M]$. From this, they obtain a relative notion of M-*cuspidal* irreducible representations for G. Harish-Chandra's theory is then recovered as the special case where G is a finite group of Lie type and M is a canonical monoid [96]. See also §8.3 of [66].

15.2 Finite Semigroups and Highest Weight Categories

Recall from § 2.3.1 that a monoid S is **regular** if, for any $x \in S$, there exists $a \in S$ such that $xax = x$. For such a semigroup, it is easy to see that there is an idempotent in each $\mathcal{J}$-class of S. In fact, x is in the $\mathcal{J}$-class of the idempotent xa. Thus,

$$S = \bigsqcup_{J \in \mathcal{U}(S)} J.$$

This may seem like a facile curiosity, imitating the much studied situation in ring theory (the one attributed to von Neumann). However, there is something of greater significance here, which was first discovered by Putcha [91]. It turns out that the complex monoid algebra of a finite regular monoid is a *quasihereditary* algebra.

Definition 15.4. *Let A be a finite dimensional algebra over K. Then A is* quasihereditary *if there is a chain of ideals (called a* hereditary chain*)*

$$0 = I_0 \subseteq I_1 \subseteq \cdots \subseteq I_r = A$$

such that, for $j = 1, \ldots, r$,

a) I_j / I_{j-1} is a projective A/I_{j-1}-module,
b) $I_j^2 = I_j$,
c) $I_j \mathcal{R} I_j \subseteq I_{j-1}$, where $\mathcal{R}$ is the radical of A.

It turns out that this is equivalent to the property that *Rep*(A) is a *highest weight category* (see Theorem 9.10 for a good illustration, and [16] for the seminal paper). In the remainder of this section we indicate how to construct a hereditary chain of ideals for the monoid algebra of a finite regular semigroup. We also show how this brings about the corresponding structure of a highest weight category. To define a highest weight category one needs simple objects, standard objects and a weight poset.

Then let M be a finite regular monoid, and let $\mathcal{U}$ denote the poset of (regular) $\mathcal{J}$-classes of M. If $e \in E(J)$, we write H_J for the $\mathcal{H}$-class of e. By the theory of Munn-Ponizovskii [15], the set $Irr(M)$ of irreducible complex representations of M is in one-to-one correspondence with the set of irreducible representations of the various H_J:

$$Irr(M) = \bigsqcup_{J \in \mathcal{U}} Irr(H_J).$$

This identification is obtained as follows. Let $L = L(\theta)$ be an irreducible complex M-module. Then there exists a unique $\mathcal{J}$-class $J \in \mathcal{U}$ such that

a) $J \cdot L \neq 0$,
b) $J' \cdot L = 0$ for all $J' \ngeq J$.

If $e \in E(J)$ is an idempotent, then $e \cdot L$ is an irreducible H_J-module. We say that

$$\theta < \theta'$$

if $J' \subseteq MJM$. Then $(Irr(M), <)$ is the weight poset of simple objects. Now for each $\theta \in Irr(H_J)$ there is a primitive idempotent $e_\theta \in \mathbb{C}[M]$ such that $e_\theta < e_J$ and $L(\theta) \cong \mathbb{C}[M]e_\theta/\mathcal{R}e_\theta$, where $\mathcal{R}$ is the radical of $\mathbb{C}[M]$. Define

$$\Delta(\theta) = \mathbb{C}[J]e_\theta.$$

Here $(\{\Delta(\theta) | \theta \in Irr(M)\}$ is the set of standard objects. The following theorem is due to Putcha (Theorem 2.1 of [91]).

Theorem 15.5. *Let M be a finite regular monoid.*

a) Then the category of finite-dimensional $\mathbb{C}[M]$-modules is a highest weight category.
b) $\mathbb{C}[M]$ is a quasihereditary algebra.

The proof of Theorem 15.5 amounts to constructing a hereditary chain as in Definition 15.4. But it turns out that there is a canonical such chain. Let $U_1 = \{J \mid \text{the minimal } \mathcal{J}\text{-class}\}$, $U_2 = \{J \mid J \text{ covers } J_1\}$, and so on, so that

$$U_m = \{J \mid J \text{ covers some } K \text{ in } U_{m-1}\}.$$

Thus define

$$J_1 = \cup_{J \in U_1} J$$
$$J_2 = \cup_{J \in U_2} J$$
$$J_3 = \cup_{J \in U_3} J$$
$$\vdots$$

Let $I_i = \mathbb{C}[J_i]$, a two-sided ideal of $\mathbb{C}[M]$, so that $I_r = \mathbb{C}[M]$ for some $r > 0$. Consider

$$0 \subseteq I_1 \subseteq I_2 \subseteq \cdots \subseteq I_r.$$

It turns out that this is the desired (canonical) hereditary chain.

15.3 Singularities of G-embeddings

It has been known for some time that any normal, spherical variety X has rational singularities in characteristic zero. The essential idea here is contained in [76], where the proof is given for the affine case. The result for general spherical varieties then follows easily from the affine case using basic local-global arguements from [13] or [36].

Furthermore, one can show that the reduction mod p of X is Frobenius split for large p. This is the best known method for obtaining geometric results about singularities in characteristic $p > 0$.

Definition 15.6. *Assume that* $charK = p > 0$*. If* X *is an algebraic variety over* K*, denote by* $F : \mathcal{O}_X \to F_*\mathcal{O}_X$ *the absolute Frobenius morphism of* X*. We say that* X *is* Frobenius split *if there is a map* $\sigma \in Hom_{\mathcal{O}_X}(F_*\mathcal{O}_X, \mathcal{O}_X)$ *such that* $\sigma \circ F$ *is the identity on* $\mathcal{O}_X$*.*

Rittatore obtains the following result in [122].

Theorem 15.7. *Any normal* G*-embedding* X *is Frobenius split.*

By a G*-embedding* we mean a $G \times G$-variety, (X, x) with dense orbit $(G \times G) \cdot x$ such that $(G \times G)_x = \{(g, h) \in G \times G \mid g = h\}$.

Corollary 15.8. *Any normal* G*-embedding* X *has rational singularities. In particular, it is a Cohen-Macaulay variety.*

Rittatore derives Theorem 15.7, first for the "wonderful" G-embedding and then derives the general case by using resolution of singularities. Notice that any reductive monoid is a G-embedding. Furthermore, the wonderful G-embedding is directly related to the canonical monoids ([96]) with group G.

In the other direction, Popov has shown in [77], that the Kraft-Procesi conjecture (to the effect that the normalization of any reductive group orbit closure has rational singularities, [50]) is false. He shows that such singularities may even be non-Cohen-Macaulay.

15.4 Cohomology of G-embeddings

Let X be a complete, nonsingular G-embedding over $\mathbb{C}$. Assume also that there is a birational $G \times G$ morphism from X to the canonical compactification of G. Assume also that $K = \mathbb{C}$, the complex numbers. Let $K \subseteq G$ be a maximal, compact subgroup in the classical topology, and let $Z \subseteq X$ be the closure of a maximal torus $T \subseteq G$. Assume that $T_K \subseteq T \cap K \subseteq K$ is a maximal torus of K. Since $KTK = G$ and KZK is compact, we obtain

$$X = KZK.$$

Thus, we obtain a surjective, real analytic morphism

$$\pi : (K \times K) \times_{T_k \times T_K} Z \longrightarrow X$$

since Z is $T_K \times T_K$-stable. Let

$$\mathfrak{U} = (K \times K) \times_{T_k \times T_K} Z \,.$$

The Weyl group W acts naturally on $\mathfrak{U}$, and π is constant on the W-orbits. Thus, W acts on $H^*(\mathfrak{U};\mathbb{Q})$ and we obtain a ring homomorphism

$$\pi^* : H^*(X;\mathbb{Q}) \longrightarrow H^*(\mathfrak{U};\mathbb{Q})^W .$$

See [22] for details.

Theorem 15.9. $\pi^* : H^*(X;\mathbb{Q}) \longrightarrow H^*(\mathfrak{U};\mathbb{Q})^W$ *is an isomorphism of rings.*

Theorem 15.9 is due to DeConcini and Procesi [22]. They also compute $H^*(\mathfrak{U};\mathbb{Q})$ in [22] by extending some work of Danilov [19]. Notice that there is a fibration

$$\mathfrak{U} \longrightarrow K/T_K \times K/T_K$$

with fibre Z. Thus $\mathfrak{U}$ may be regarded as a "relative" torus embedding.

In [5] these same authors (along with Bifet) describe the rational cohomology ring of any complete symmetric variety by generators and relators. The authors first introduce a more general notion. A **regular embedding** is a smooth algebraic variety X on which a connected affine algebraic group G acts with finitely many orbits. Every orbit closure is smooth and is the transverse intersection of the codimension one orbit closures which contain it. Moreover, the isotropy group of any $x \in X$ must have a dense orbit in the normal space to the orbit $G \cdot x \subseteq X$. Examples of projective regular embeddings can often be obtained from semisimple algebraic monoids M with zero, such that $M\backslash\{0\}$ is smooth. The authors then determine the G-equivariant rational cohomology algebra $H_G^*(X)$ of any regular embedding X, by associating with X a "Stanley-Reisner system" (a generalization of the Stanley-Reisner algebra associated with a simplicial complex). Using this construction, they construct an algebra which turns out to be $H_G^*(X)$. If X is also compact, the cohomology ring $H^*(X)$ can be obtained as the quotient of $H_G^*(X)$ by a certain regular sequence consisting of generators of $H_G^*(\{*\})$, the one point G-space. For X a torus embedding, one recovers the Jurkiewicz-Danilov presentation of $H^*(X)$ as discussed by Danilov in [19].

15.5 Horospherical Varieties

Definition 15.10. *An embedding X of G/H is called* horospherical *if H contains a maximal unipotent subgroup of G.*

Horospherical embeddings were first looked at systematically by Popov and Vinberg in [79]. They obtained the fundamental correspondence between the poset of orbits and the poset of faces of the corresponding polyhedral cone. They also characterized normal and factorial horospherical embeddings. The results of [78] and [79] were cited by Luna and Vust (in their seminal paper [56]) as a major catalyst in the development of the theory of spherical embeddings. A partial generalization to the nonaffine case was carried out by Pauer in [70, 71].

The technique of "horospherical degeneration" [76] has since become an important method in the study of spherical varieties [1, 142].

15.6 Monoids associated with Kac-Moody Groups

In [43] Kac and Peterson start with a Kac-Moody Lie algebra $\mathfrak{g}$ and construct a group G, the associated Kac-Moody group. In [50] they introduce an algebra, the algebra of **strongly regular functions** on G.

Let $\mathbb{F}$ be an algebraically closed field of characteristic zero, and let $\mathfrak{g}$ be a Lie algebra defined over $\mathbb{F}$. Assuming that $\mathfrak{g}$ comes from a symmetrizable, generalized Cartan matrix, they define (in [50]) the algebra $\mathbb{F}[G]$ of *strongly regular functions* on G. They show that

$$\mathbb{F}[G] = \bigoplus_{\lambda \in P^+} L^*(\lambda) \otimes L(\lambda)$$

where P^+ is the set of dominant weights of the weight lattice P. Since each $L(\Lambda)$ is integrable they obtain

$$j : G \subseteq \operatorname{Specm}(\mathbb{F}[G]),$$

the space of codimension one ideals of $\mathbb{F}[G]$. However, j is not surjective unless $\mathfrak{g}$ is finite dimensional.

About the same time (1983), Slodowy [125] suggested that there should exist a some kind of partial compactification $\overline{G} \subseteq \operatorname{Specm}(\mathbb{F}[G])$ whose structure might significantly help in the study of the deformation of certain singularities. He conjectured that

$$\overline{G} = G\overline{T}G$$

and he also anticipated the structure of $\overline{T}$. Peterson suggested that $\overline{G}$ might actually be a monoid.

Since then, Mokler [58, 59] approached these problems systematically. He first constructs a monoid

$$\widehat{G} \subseteq End\left(\bigoplus_{\Lambda \in P^+} L(\Lambda)\right)$$

directly, using the set of faces $\mathfrak{F} = \{R \subseteq X\}$ of the **Tits cone** X, to define projections

$$e(R) : \bigoplus_{\Lambda \in P^+} L(\Lambda) \longrightarrow \bigoplus_{\Lambda \in P^+} L(\Lambda).$$

$\widehat{G}$ is then the submonoid of $End\left(\bigoplus_{\Lambda \in P^+} L(\Lambda)\right)$ generated by $G \cup \{e(R) \mid R \in \mathfrak{F}\}$. He then proves [58] that

$$\widehat{G} = \overline{G} \subseteq End(\oplus L(\Lambda))$$

the closure being taken in the sense of elementary, infinite dimensional, algebraic geometry (section 3 of [58]). He also obtains some decisive structural information about $\overline{G}$.

Induced by its action on X, the Weyl group $\mathcal{W}$ acts on the face lattice $\mathfrak{F}$ of X. A face $R = R(\Theta)$ of X is called **special** if the relative interior of R meets the closed fundamental chamber $\overline{C}$ of X. (The situation is similar to what we encountered in § 7.2.) Let I be the set of vertices of the underlying Dynkin diagam of G. Then each special face of X corresponds to a subset of I for which the associated Dynkin subdiagram is either empty, or has only components of infinite type.

Theorem 15.11.

$$\widehat{G} = \bigsqcup_{R \text{ special}} Ge(R)G\,.$$

For the proof see Proposition 2.26 of [58].

Mokler continues his work in [59] by further investigating the structure of $Specm(\mathbb{F}[G])$. In [60], he obtains results generalizing the work of [105, 112, 73]. In particular, he determines the generalized Bruhat decomposition on $\widehat{M}$ and the length function on $\widehat{W}$.

References

1. I. V. Arzhantsev, *Contractions of affine spherical manifolds*, Mat. Sb. 190(1999), 3-22.
2. M. F. Atiyah and I. G. MacDonald, *"Commutative algebra"*, Addison-Wesley, London, 1969.
3. M. K. Augustine, *Congruences of monoids of Lie type* In *"Monoids and semigroups with applications"*, held at Berkeley, 1989, 324-333, World Sci. Publishing, River Edge, NJ, 1991.
4. A. Bialynicki-Birula, *Some theorems on actions of algebraic groups*, Annals of Math. 98(1973), 480-497.
5. E. Bifet, C. De Concini, C. Procesi, *Cohomology of regular embeddings*, Advances in Math. 82(1990), 1-34.
6. A. Björner, M. Wachs, *Bruhat order of Coxeter groups and shellability*, Advances in Math. 43(1982), 87-100.
7. A. Borel, *Linear algebraic groups*, Graduate Texts in Mathematics 126, Springer-Verlag, New York, 1991.
8. D. Borwein, S. Rankin, L. Renner, *Enumeration of injective partial transformations*, Discrete Math. 73(1989), 291-296.
9. M. Brion, *Quelques propriétés des espaces homogènes sphériques*, Manuscripta Math. 55(1986), 191-198.
10. M. Brion, *Sur la géométrie des variétés sphériques*, Comment. Math. Helv. 66(1991) 237-262.
11. M. Brion, *Variétés sphériques*, In, *"Opérations hamiltoniennes et opérations de groupes algébriques"*, Notes de la session de S. M. F. Grenoble, 1997.
12. M. Brion, *The behaviour at infinity of the Bruhat decomposition*, Comment. Math. Helv. 73(1998), 137-174.
13. M. Brion, D. Luna, Th. Vust, *Espace homogène sphérique*, Invent. Math. 84(1986), 617-632.
14. C. Chevalley, *Sur certain groupes simple*, Tôhoku Math. J. 7(1955),14-66.
15. A. H. Clifford, G. B. Preston, *Algebraic theory of semigroups, Vol.1*, AMS Surveys Vol. 7, Providence, Rhode Island, 1961.
16. E. Cline, B. Parshall, L. Scott, *Finite dimensional algebras and highest weight categories*, J. Reine Angew. Math. 391(1988), 85-99.
17. D. A. Cox, *The homogeneous coordinate ring of a toric variety*, J. Algebraic Geometry 4(1995), 17-50.

18. C. W. Curtis, *Modular representations of finite groups with split BN pair*, Springer Lecture Notes 131(1970), 57-95.
19. V. I. Danilov, *The geometry of toric varieties*, Russian Math. Surveys 33(1978), 97-154.
20. C. De Concini, *Normality and nonnormality of of certain semigroups and orbit closures*, in: *"Interesting algebraic varieties arising in algebraic transformation groups"*, Encyclopedia of Mathematical Sciences vol.132, Subseries: Invariant Theory and Algebraic Transformation Groups, vol. III, Springer-Verlag, 2004, 15–35.
21. C. De Concini, C. Procesi *Complete symmetric varieties*, Springer Lecture Notes in Mathematics 131(1983), 1-44.
22. C. De Concini, C. Procesi, *Cohomology of compactifications of algebraic groups*, Duke Math. Journal 53(1986), 585-594.
23. S. Donkin, *The blocks of a semisimple algebraic group*, Journal of Algebra 67(1980), 36-53.
24. S. Donkin, *Good filtrations of rational modules for reductive groups*, Proc. Symp. Pure Math. 47(1987), 69-80.
25. S. Donkin, *On Schur algebras and related algebras IV. The blocks of the Schur algebra*, J. of Alg. 168(1994), 400-429.
26. S. Doty, *Representation theory of reductive normal monoids*, Trans. Amer. Math. Soc. 351(1999), 2539-2551.
27. R. R. Douglas, L. E. Renner, *Uniqueness of product and coproduct decompositions in rational homotopy theory*, Trans. Amer. Math. Soc. 264(1981), 165-180.
28. E. Elizondo, K. Kurano, K. Watanabe, The total coordinate ring of a normal projective variety, Journal of Algebra 276(2004), 625-637.
29. R. Fossum, *"The divisor class group of a Krull domain"*, Springer Verlag, New York, 1973.
30. W. Fulton, *"Intersection theory"*, Springer Verlag, New York, 1984.
31. W. Fulton, *"Introduction to Toric Varieties"*, Princeton University Press, New Jersey, 1993.
32. J. A. Green, *On the structure of semigroups*, Annals of Math. 54(1951), 163-172.
33. J. A. Green, *Locally finite representations*, Journal of Algebra 41(1976), 137-171.
34. F. D. Grosshans, *Observable groups and Hilbert's fourteenth problem*, Amer. J. Math. 41(1976), 229-253.
35. F. D. Grosshans, *Contractions of the actions of reductive groups in arbitrary characteristic*, Inv. Math. 107(1992),127-133.
36. F. D. Grosshans, *Constructing invariant polynomials via Tschirnhaus transformations*, Lecture Notes in Math. vol. 1278, Springer-Verlag, 1987.
37. H. Hiller, *"Geometry of Coxeter groups"*, Pitman Advanced Publishing Program, Research Notes in Mathematics vol. 54, Boston, 1982.
38. R. Hartshorne, *"Algebraic geometry"*, Graduate Texts in Mathematics 52, Springer-Verlag, New York, 1977.
39. W. Huang, *Reductive and semisimple algebraic monoids*, Forum Mathematicum 13(2001), 495-504.
40. J. E. Humphreys, *"Linear algebraic groups"*, 3rd Ed., Graduate Texts in Mathematics 21 Springer-Verlag, New York, 1987.

41. J. E. Humphreys, *"Reflection groups and Coxeter groups"*, Cambridge University Press, Cambridge, 1990.
42. B. Iversen, *The geometry of algebraic groups*, Advances in Math. 20(1976), 57-85.
43. V. Kac, D. Peterson, *Infinite flag varieties and conjugacy theorems*, Proc. Nat. Acad. Sci. 80(1983), 1778-1782.
44. V. Kac, D. Peterson, *Regular functions on certain infinite dimensional groups*, Progress in Math. 36, Birkhauser, Boston, 1983, 141-166.
45. S. S. Kannan, *On the projective normality of wonderful compactifications of semisimple adjoint groups*, Math. Z. 239(2002), 673-682.
46. S. Kato, *A Borel-Weil-Bott type theorem for group completions*, Journal of Algebra 259(2003), 572-580.
47. G. Kempf, F. Knudsen, D. Mumford, B. Saint-Donat, *"Toroidal embeddings I"*, Springer Verlag, New York, 1973.
48. F. Knop, *The Luna-Vust theory of spherical varieties*, "Proceedings of the Hyderabad Conference Conference on Algebraic Groups", S. Ramanan et. al. Eds., NBHM, Manoj, 1991, 225-249.
49. L. Kovacs, *Semigroup algebras of the full matrix group over a finite field*, Proc. Amer. Math. Soc. 116(1992), 911-919.
50. H. Kraft, C. Procesi, *On the geometry of conjugacy classes in classical groups*, Comm. Math. Helv. 57(1982), 539-602.
51. Z. Li, *The lattice of $\mathfrak{J}$-classes of $(\mathfrak{J},\sigma)$-irreducible monoids II*, Journal of Algebra 221(1999), 117-134.
52. Z. Li, M. S. Putcha, *Types of reductive monoids* Journal of Algebra 221(1999), 102-116.
53. Z. Li and L. E. Renner, *The lattice of $\mathfrak{J}$-classes of $(\mathfrak{J},\sigma)$-irreducible monoids*, Journal of Algebra 190(1997), 172-194.
54. P. Littelmann, C. Procesi, *Equivariant cohomology of wonderful compactifications*, "Operator algebras, unitary representations, enveloping algebras, and invariant theory", 219-262, Progr. Math., 92, Birkhuser, Boston, MA, 1990.
55. D. Luna, *Report on spherical varieties*, notes, 1086.
56. D. Luna, Th. Vust, *Plongements d'espaces homogènes*, Comment. Math. Helv. 58 (1983), 186-245.
57. H. Matsumura, *"Commutative algebra"*, Benjamin/Cummings, Reading Mass., 1980.
58. C. Mokler, *An analogue of a reductive algebraic monoid, whose unit group is a Kac-Moody group*, Mem. of the Amer. Math. Soc., to appear, submitted 2003.
59. C. Mokler, *The $\mathbb{F}$-valued points of the algebra of strongly regular functions of a Kac-Moody group*, Transformation Groups 7(2002), 343-378.
60. C. Mokler, *Extending the Bruhat order and the length function from the Weyl group to the Weyl monoid*, Journal of Algebra 275(2004), 815-855.
61. D. Mumford, *"The red book of varieties and schemes"*, Springer Lecture Notes 1358, New York, 1988.
62. D. Mumford, J. Fogarty, F. Kirwan, *"Geometric invariant theory"*, Springer Verlag, 1994.
63. W. D. Munn, *Pseudo-inverses in semigroups*, Proc. Cam. Phil. Soc. 57(1961), 247-250.
64. K. S. S. Nambooripad, *"Structure of regular semigroups I"*, Mem. Amer. Math. Soc. 224(1979).

65. P. E. Newstead, *"Introduction to moduli problems and orbit spaces"*, Tata Institute Lectures on Mathematics, 51, Bombay, by the Narosa Publishing House, New Delhi, 1978.
66. J. Okñinski, *"Semigroups of matrices"*, Series in Algebra vol. 6, World Scientific, 1998.
67. J. Okñinski, M. S. Putcha, *Complex representations of matrix semigroups*, Trans. Amer. Math. Soc. 323(1991), 563-581.
68. J. Okñinski, M. S. Putcha, *Parabolic subgroups and cuspidal representations of finite monoids*, Int. J. Alg. Comp. 1(1991), 33-47.
69. A. Onishchik and E. Vinberg, *"Lie groups and algebraic groups"*, Springer Verlag, 1990.
70. F. Pauer, *Normale Einbettungen von G/U*, Math. Ann. 257(1981), 371-396.
71. F. Pauer, *Glatte Einbettungen von G/U*, Math. Ann. 262(1983), 421-429.
72. E. A. Pennell, *Generalized Bruhat order on reductive monoids*, Ph. D. Thesis, North Carolina State University, Raleigh, 1995.
73. E. A. Pennell, M. S. Putcha, L. E. Renner, *Analogue of the Bruhat-Chevalley order for reductive monoids*, Journal of Algebra 196(1997), 339-368.
74. V. L. Popov, *Stability criterion for the action of a semisimple group on a factorial manifold*, Izv. Akad. Nauk. USSR 34(1973), 527-535.
75. V. L. Popov, *Picard groups of homogeneous spaces of linear algebraic groups and one-dimensional homogeneous vector bundles*, Math. USSR-Izv. 8(1974), 301-327.
76. V. L. Popov, *Contraction of the actions of reductive algebraic groups*, Math. USSR Sbornik, 58(1987), 311-368.
77. V. L. Popov, *Singularities of closures of orbits*, Israel Math. Conf. Proc. vol. 7, Amer. Math. Soc., 1993, 133–141
78. V. L. Popov, *Quasihomogeneous affine algebraic varieties of the group SL(2)*, Math. USSR-Izv. 4(1973), 793-831.
79. V. L. Popov, E. B. Vinberg, *On a class of quasihomogeneous affine varieties*, Math. USSR-Izv. 6(1972), 743-758.
80. R. A. Proctor, *Classical Bruhat orders and lexicographic shallabillity*, Journal of Algebra 77(1982), 104-126.
81. M. S. Putcha, *Green's relations on a connected algebraic monoid*, Linear and Multilinear Algebra 12(1982), 37-50.
82. M. S. Putcha, *"Linear algebraic monoids"*, Cambridge University Press, 1988.
83. M. S. Putcha, *Determinant functions on algebraic monoids*, Comm. in Alg. 11(1983), 695-710.
84. M. S. Putcha, *Regular linear algebraic monoids*, Trans. Amer. Math. Soc. 290(1985), 615-626.
85. M. S. Putcha, *Monoids on groups with BN pair*, Journal of Algebra 120(1989), 139-169.
86. M. S. Putcha, *Classification of monoids of Lie type*, Journal of Algebra, 120(1994), 636-662.
87. M. S. Putcha, *Monoid Hecke algebras*, Trans. Amer Math. Soc. 349(1997), 3517-3534.
88. M. S. Putcha, *Conjugacy classes in algebraic monoids*, Teans. Amer. Math. Soc. 303(1987), 529-540.
89. M. S. Putcha, *Conjugacy classes in algebraic monoids II*, Can. J. of Math. 46(1994), 648-661.

90. M. S. Putcha, *Conjugacy classes and nilpotent variety of a reductive monoid*, Can. J. of Math. 50(1998), 829-844.
91. M. S. Putcha, *Complex representations of finite monoids. II. Highest weight categories and quivers*, Journal of Algebra 205(1998), 53-76.
92. M. S. Putcha, *Shellability in reductive monoids*, Trans. Amer. Math. Soc. 354(2002), 413-426.
93. M. S. Putcha, *Bruhat-Chevalley order in reductive monoids*, to appear.
94. M. S. Putcha, *Shellability in reductive monoids II*, to appear.
95. M. S. Putcha, L. E. Renner, *The system of idempotents and lattice of J-classes of reductive algebraic monoids*, Journal of Algebra 116(1988), 385-399.
96. M. S. Putcha, L. E. Renner, *The canonical compactification of a finite group of Lie type*, Trans. Amer. Math. Soc. 337(1993), 305-319.
97. M. S. Putcha, L. E. Renner, *Morphisms and duality of monoids of Lie type*, Journal of Algebra 184(1996), 1025-1040.
98. M. S. Putcha, L. E. Renner, *The orbit structure of 2-reducible algebraic monoids*, J. Alg. Comb., to appear.
99. L. E. Renner, *"Automorphism groups of minimal algebras"* UBC Thesis, 1978.
100. L. E. Renner, *"Algebraic monoids"*, UBC Thesis, Vancouver, 1982.
101. L. E. Renner, *Quasi-affine algebraic monoids*, Semigroup Forum, 30(1984), 167-176.
102. L. E. Renner, *Reductive monoids are von Neumann regular*, Journal of Algebra 93(1985), 237-245.
103. L. E. Renner, *Classification of semisimple rank-one monoids*, Trans. Amer. Math. Soc. 287(1985), 457-473.
104. L. E. Renner, *Classification of semisimple algebraic monoids*, Trans. Amer. Math. Soc. 292(1985), 193-223.
105. L. E. Renner, *Analogue of the Bruhat decomposition for algebraic monoids*, Journal of Algebra 101(1986), 303-338.
106. L. E. Renner, *Connected algebraic monoids*, Semigroup Forum 36(1987),365-369.
107. L. E. Renner, *Conjugacy classes of semisimple elements and irreducible representations of algebraic monoids*, Comm. in Alg. 16(1988), 1933-1943.
108. L. E. Renner, *Completely regular algebraic monoids*, J. Pure and Appl. Alg. 59(1989), 291-298.
109. L. E. Renner, *Classification of semisimple varieties*, Journal of Algebra 122(1989), 275-287.
110. L. E. Renner, *Finite monoids of Lie type*, in *"Monoids and semigroups with applications"*, J. Rhodes Ed., World Scientific(1991), 278-287.
111. L. E. Renner, *The homotopy types of retracts of a fixed space*, J. Pure Appl. Alg. 69(1991), 295-299.
112. L. E. Renner, *Analogue of the Bruhat decomposition for algebraic monoids II: the length function and the trichotomy*, Journal of Algebra 175(1995), 697-714.
113. L. E. Renner, *The blocks of solvable algebraic monoids*, Journal of Algebra 188(1997), 272-291.
114. L. E. Renner, *Modular representations of finite monoids of Lie type*, J. of Pure and Appl. Alg. 138(1999), 279-296.
115. L. E. Renner, *Distribution of products in finite monoids I: Combinatorics*, International Journal of Algebra and Computation 9(1999), 693-708.

116. L. E. Renner, *Distribution of products in finite monoids II: Algebra*, International Journal of Algebra and Computation 9(1999), 709-720.
117. L. E. Renner, *Regular algebraic monoids*, Semigroup Forum 63(2001), 107-113.
118. L. E. Renner, *An explicit cell decomposition of the canonical compactification of an algebraic group*, Can. Math. Bull., 46(2003), 140-148.
119. J. Rhodes, *Characters and complexity of finite semigroups*, J. of Comb. Theory 6(1969), 67-85.
120. A. Rittatore, *Monoid algébrique et plongemont des groupes*, Thesis de Université Joseph Fourier, Grenoble, 1997.
121. A. Rittatore, *Algebraic monoids and group embeddings*, Transformation Groups 3(1998), 375-396.
122. A. Rittatore, *Reductive embeddings are Cohen-Macaulay*, Proc. Amer. Math. Soc. 131(2003), 675-684.
123. A. Rittatore, *Very flat reductive monoids*, Publ. Mat. Urug. 9(2001), 93-121.
124. J. P. Serre, *"Linear representations of finite groups"*, Springer Verlag, New York, 1977.
125. P. Slodowy, private communication, 1984.
126. S. Smith, *Irreducible modules and parabolic subgroups*, Journal of Algebra 75(1982), 286-289.
127. L. Solomon, *The Bruhat decomposition, Tits system, and Iwahori ring for the monoid of matrices over a finite field*, Geom. Dedicata 36(1990), 15-49.
128. L. Solomon, *An introduction to reductive monoids*, "Semigroups, formal languages and groups", J. Fountain, Ed., Kluwer Academic Publishers, 1995, 295-352.
129. L. Solomon, *Representations of the rook monoid*, Journal of Algebra 256(2002), 309-342.
130. T. A. Springer, *Intersection cohomology of $B \times B$-orbit closures in group compactifications*, Journal of Algebra 258(2002), 71-111.
131. T. A. Springer, R. Steinberg, *Conjugacy classes*, Springer Lecture Notes 131(1970), 167-266.
132. R. P. Stanley, *"Combinatorics and commutive algebra"*, (2nd Ed.), Birkhäuser, New York, 1996.
133. R. Steinberg, *"Endomorphisms of linear algebraic groups"*, Mem. Amer. Math. Soc. 80(1968)
134. R. Steinberg, *"Conjugacy classes in algebraic groups"*, Springer Lecture Notes 366, New York, 1974.
135. R. Steinberg, *Variations on a theme of Chevalley*, Pac. J. Math. 9(1959), 875-891.
136. R. Steinberg, *Regular elements of semisimple algebraic groups*, Inst. des Hautes Etudes Sci. Publ. Math. 25(1965), 49-80.
137. R. Steinberg, *Torsion in reductive groups*, Adv. in Math. 15(1975), 63-92.
138. D. A. Timashev, *Equivariant compactifications of reductive groups*, Sbornik: Mathematics 194 (2003), 589-616.
139. D. A. Timashev, *"Spherical embeddings"*, EMS series, Springer-Verlag, in preparation.
140. J. Tits, *Abstract and algebraic simple groups*, Annals of Math. 80(1964), 313-329.
141. È. B. Vinberg, *Complexity of actions of reductive groups*, Functional Anal. Appl. 20(1986), 1-13.

142. È. B. Vinberg, *On reductive algebraic semigroups*, Amer. Math. Soc. Transl., Series 2, 169(1994), 145-182.
143. È. B. Vinberg, *The asymptotic semigroup of a semisimple Lie group* in "Semigroups in algebra, geometry and analysis" 293-310, de Gruyter Exp. Math., 20, de Gruyter, Berlin, 1995.
144. Th. Vust, *Opération de groupes réductif dans un type de cône presque homogènes*, Bull. Math Soc. France 102(1974), 317-334.
145. W. C. Waterhouse, *The unit groups of affine algebraic monoids*, Proc. Amer. Math. Soc. 85(1982), 506-508.

Index

Printing: Krips bv, Meppel
Binding: Litges & Dopf, Heppenheim